Mechanical Engineering Series

Series Editor
Francis A. Kulacki
Department of Mechanical Engineering, University of Minnesota,
Minneapolis, MN, USA

The Mechanical Engineering Series presents advanced level treatment of topics on the cutting edge of mechanical engineering. Designed for use by students, researchers and practicing engineers, the series presents modern developments in mechanical engineering and its innovative applications in applied mechanics, bio-engineering, dynamic systems and control, energy, energy conversion and energy systems, fluid mechanics and fluid machinery, heat and mass transfer, manufacturing science and technology, mechanical design, mechanics of materials, micro- and nano-science technology, thermal physics, tribology, and vibration and acoustics. The series features graduate-level texts, professional books, and research mono-graphs in key engineering science concentrations.

More information about this series at http://www.springer.com/series/1161

Verica Radisavljević-Gajić • Miloš Milanović •
Patrick Rose

Multi-Stage and Multi-Time Scale Feedback Control of Linear Systems with Applications to Fuel Cells

Springer

Verica Radisavljević-Gajić
Department of Mechanical Engineering
Villanova University
Villanova, PA, USA

Miloš Milanović
Department of Mechanical Engineering
Villanova University
Villanova, PA, USA

Patrick Rose
Department of Mechanical Engineering
Villanova University
Villanova, PA, USA

ISSN 0941-5122 ISSN 2192-063X (electronic)
Mechanical Engineering Series
ISBN 978-3-030-10388-0 ISBN 978-3-030-10389-7 (eBook)
https://doi.org/10.1007/978-3-030-10389-7

Library of Congress Control Number: 2018966525

This Springer imprint is published by the registered company Springer Nature Switzerland AG
The registered company address is: Gewerbestrasse 11, 6330 Cham, Switzerland

"I dedicate this book to my parents
Ljiljana and Miroljub Radisavljević
for their unconditional love and support"
Dr. Verica Radisavljević-Gajić

"To my dear family and Saša,
for endless support and lifelong inspiration"
Miloš Milanović

"To my friends and family
for their endless support and confidence
in me"
Patrick Rose

Preface

This monograph is intended for researchers and practitioners in control systems interested in complex, time-invariant, linear dynamic systems composed of several subsystems and/or large-scale linear, time-invariant, dynamic systems operating in several time scales. It can be used by engineering graduate students majoring in control systems and automation, practicing control engineers, as well as all engineering, applied mathematics, economics, and computer science faculty interested in control of dynamic systems. Due to numerous applications of the derived feedback controller design algorithms presented in this book to fuel cells (clean electric energy sources), particularly to proton-exchange membrane (PEM) fuel cells, the book is also of interest for the fuel cell community researchers, graduate students, and practitioners. One of the most important features of the presented design methodology is that different types of linear feedback controllers can be designed for different subsystems and/or in different time scales of a complex linear dynamic system.

This research monograph presents completely the *two-stage* feedback controller design algorithms in both *continuous-* and *discrete-tim*e domains including all design formulas and algebraic equations for general linear time-invariant dynamic systems. The results presented are specialized and simplified for the *two-time scale* linear time-invariant dynamic systems (singularly perturbed systems). The corresponding presentation is completely done also for the *three-stage continuous-time* feedback controller designs for *general* linear time-invariant dynamic systems. Extensions of three-stage feedback controllers to linear dynamic time-invariant continuous- and discrete-time systems operating in the three-time scales have been done only in the continuous-time. The *discrete-time three-time scale* systems remain an open research area for the three-stage design of linear feedback controllers, mostly due to unsettled issues about the formulations of the three-time scale linear dynamic systems. The *four-stage* linear feedback controller design is presented only for *general* class of linear, time-invariant, continuous-time dynamic systems. How to extend those results or, more precisely, how to solve the obtained design algebraic equations for the four-time scale, time-invariant, continuous-time, linear dynamic systems is discussed. Hence, the *four-stage four-time* scale feedback controller design is left as an open future challenging research area.

The designs of linear-quadratic optimal controllers and eigenvalue assignment ("pole placement") feedback controllers are fully covered in this research monograph. The book provides opportunities for future research of other types of multistage linear feedback controllers for general and multi-time scale linear dynamic systems, including observer-based and Kalman filter-based deterministic and stochastic controllers. Due to the fact that different types of linear feedback controllers can be designed (partial controllers) for different subsystems of a large-scale linear dynamic system and/or system operating in different time scales, producing the dual results for observers and the Kalman filter, it will be possible, for example, to design partially optimal Kalman filter and new classes of reduced-order observers. Extensions to other types of existing linear feedback controllers and filters are possible, including the design of new hybrid linear dynamic feedback controllers and filters.

Villanova, PA, USA Verica Radisavljević-Gajić
Villanova, PA, USA Miloš Milanović
Villanova, PA, USA Patrick Rose

Contents

Chapter 1
Introduction

The design of linear continuous- and discrete-time state feedback controllers is well documented in control engineering literature; see, for example, Franklin et al. (1990), Ogata (1995), Sinha (2007), and Chen (2012). The authors of this monograph have recently developed new algorithms for the design of two- and three-stage feedback controllers for both linear discrete- and continuous-time dynamic systems (Radisavljevic-Gajic and Rose 2014; Radisavljevic-Gajic 2015a, b; Radisavljevic-Gajic et al. 2015, 2017) that have been efficiently applied to two- and three-time-scale models of fuel cells (Radisavljevic-Gajic and Rose 2014; Radisavljevic-Gajic et al. 2015, 2017; Radisavljevic-Gajic and Milanovic 2016; Milanovic et al. 2017; Milanovic and Radisavljevic-Gajic 2018). In general, the results of these new multistage and multi-time-scale feedback controller design algorithms are applicable under mild conditions to almost all linear continuous- and discrete-time time-invariant linear systems.

Fuel cells produce electricity from hydrogen-rich fuels via chemical reactions without burning fuel (Larminie and Dicks 2001; Barbir 2005; Nehrir and Wang 2009; Gou et al. 2010; Hoffmann and Dorgan 2012; Eikerling and Kulikovsky 2014). The type of fuel cells considered in this book is the proton-exchange membrane (PEM) fuel cells, also known as polymer exchange membrane fuel cells. A PEM fuel cell is a triode composed of an anode, membrane, and cathode. Hydrogen is pumped from the anode side, and oxygen is pumped from the cathode side. They are the most developed and the best understood type of fuel cells used for both mobile and stationary applications. PEM fuel cells are devices that rely on the chemical production of electric energy and water by mixing in a specific manner hydrogen and oxygen (a process reverse to water electrolysis). This process was discovered in the middle of the nineteenth century (almost 200 years ago), but due to its multidisciplinary nature complexity, it took a long time to become a mature technology. It is interesting to point out that in the 1960s within the Apollo Space.

Program, NASA used fuel cells to provide water for astronauts from the tanks of oxygen and hydrogen.

V. Radisavljević-Gajić et al., *Multi-Stage and Multi-Time Scale Feedback Control of Linear Systems with Applications to Fuel Cells*, Mechanical Engineering Series,
https://doi.org/10.1007/978-3-030-10389-7_1

Modeling, control, and simulation of PEM fuel cells have been a very active research area; see, for example, Pukrushpan et al. (2004a, b), Nehrir and Wang (2009), Gou et al. (2010), Wang et al. (2011, 2013), Barelli et al. (2012), Matraji et al. (2012, 2013, 2015), Bhargav et al. (2014), Jiao (2014), Wang and Guo (2015), Li et al. (2015a, b), Naghidokht et al. (2016), Wu and Zhou (2016), Zhou et al. (2017), Hong et al. (2017), Tong et al. (2017), Daud et al. (2017), Reddy and Samuel (2017), Majlan et al. (2018), Sankar and Jana (2018a, b), and references therein. The importance of mathematical modeling for studying fuel cell dynamics is emphasized in Fuhrmann et al. (2008): "The operation of fuel cells with polymer electrolyte membranes (PEMs) is based on complex interactions of physical, chemical, and electrochemical processes on multiple time scales. A quantitative and qualitative understanding of this complex matter is possible only on the base of mathematical models." Mathematical modeling of fuel cells should be done very carefully combining knowledge from several scientific and engineering disciplines such as mathematics, physics, chemistry, system analysis, and control engineering.

One of the most important applications of PEM fuel cells is for electric cars. A single plate PEM fuel cell, whose width is very narrow, 1 mm, produces the voltage of only 0.7 V. It has the current density of $\approx$0.8 A/cm^2 so that a plate of only 10 cm $\times$ 10 cm produces 56 W. To get much higher voltage and electric power, n fuel cell plates are connected in series to form a fuel cell stack. To drive a fuel cell-powered electric car, a nominal electric power of ~40 kW (sufficient to provide electricity for eight average houses) is needed that requires PEM fuel cell electric power of at least 80 kW, which can be provided by 228 fuel cell plates of dimension 25 cm $\times$ 25 cm connected in series. For comparison, to produce ~80 kW, Tesla battery-powered electric cars use 8000 batteries of 1 V. Note that Honda Clarity, Hyundai Tucson already on our roads, and Mercedes-Benz B-Class (still in the testing phase) use fuel PEM cells of 100 kW. Toyota Mirai (also commercially available) has a PEM fuel cell of 114 kW, as reported in Rojas et al. (2017). A very recent article by Samuelsen (2017) published in IEEE Spectrum was entitled "The automotive future belongs to fuel cells: Range, adaptability, and refueling time will ultimately put hydrogen fuel cells ahead of batteries." Several research publications study modeling and control issues of fuel cells (particularly PEM fuel cells) for automotive applications; see, for example, zur Megede (2002), Pukrushpan et al. (2004a, b), Mitchell et al. (2006), Wang and Peng (2014), Haddad et al. (2015), Reddy and Samuel (2017), Han et al. (2017), and Zhou et al. (2017, 2018).

The two- and three-stage designs of feedback controllers were simplified and specialized in Radisavljevic-Gajic and Rose (2014), Radisavljevic-Gajic (2015a, b), and Radisavljevic-Gajic et al. (2017) for two- and three-time-scale systems (also known as singularly perturbed systems) that have a natural decomposition into slow and fast subsystems, and, hence, they are very well suited for the two- and three-stage feedback controller design techniques. Dynamic systems with slow and fast state space variables play important roles in control engineering; see, for example, the books and overview papers by Kokotovic et al. (1999), Naidu and Calise (2001), Gajic and Lim (2001), Liu et al. (2003), Dimitriev and Kurina (2006), Zhang et al. (2014), Kuehn (2015), and references therein.

Several time scales are present in many real physical systems that have components of different nature (electrical, mechanical, chemical, thermodynamic, electrochemical). For example, advanced heavy water reactor (Shimjith et al. 2011a, b; Munje et al. 2014) has three time scales. Dynamics of fuel cells evolves in at least three, possibly four, time scales (Zenith and Skogestad 2009). It was shown in Zenith and Skogestad (2009) that a proton-exchange membrane fuel cell (PEMFC) system has three subsystems operating in three different time scales corresponding to three different time constants: electrochemical subsystem operating in seconds, chemical part of the PEMPC system (energy balance and mass balance) operating in minutes, and electrical part of the PEMFC system operating in milliseconds. Road vehicles possess the multi-time scale dynamics as demonstrated in Wedig (2014). Chemical reaction networks can be modeled using the multi-time-scale analysis (Lee and Othmer 2010). It is interesting to point out that the Hodgkin-Huxley equation that models nerve electric conductivity has a singularly perturbed form as reported in Cronin (2008). The paper by Jalics et al. (2010) showed that the neuron dynamic model can be studied in three time scales. In power electronics, many devices operate in three time scales (Umbria et al. 2014), and in general, power systems composed of electrical, mechanical, and electronic components possess several time scales due to the presence of several time constants of different order of magnitudes. Three time scales can be found in helicopter dynamics (Esteban et al. 2013) and *five times scales* are needed to model electron dynamics (Kummrow et al. 1999).

The power of the two- and three-stage (and in general multistage) feedback design techniques is summarized in the following:

(a) Different types of controllers (eigenvalue assignment, optimal, robust, reliable, etc.) can be designed for different parts of the system (subsystems) using corresponding feedback gains obtained by performing calculations (design) only with subsystem (reduced-order) matrices.
(b) Local subsystem feedback gains, for example, in the case of the two-stage feedback design, G_1 and G_2, which control local subsystems are compounded into one full-state feedback gain via a simple formula $G_{eq} = G_{eq}(G_1, G_2)$, leading to the standard full-state feedback controller. The corresponding block diagram for the two-stage design is presented in Fig. 1.1.
(c) Computational requirements are drastically reduced (especially for two- and three-time scale linear systems) since all numerical operations are done with matrices of the reduced-order corresponding to the subsystems.

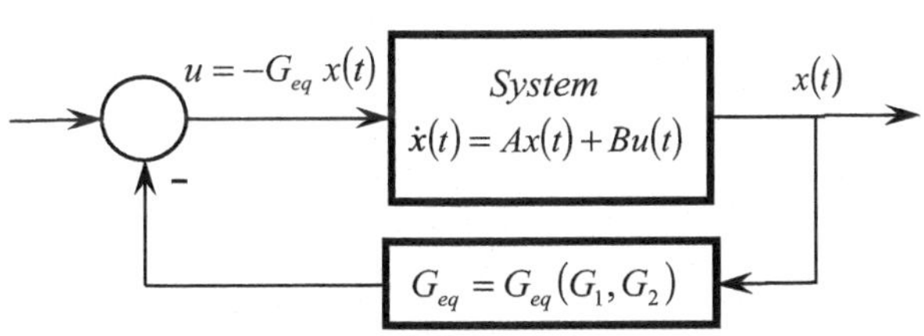

Fig. 1.1 Two-stage full-state feedback $G_{eq} = G_{eq}(G_1, G_2)$ with G_1 controlling a part of the system and G_2 controlling the rest of the system

(d) Very high accuracy can be achieved since numerical ill-conditioning of higher-order matrices can be eliminated and computations performed with well-conditioned lower-order matrices (especially for two- and three-time scale systems).
(e) The design can be extended for the development of corresponding two- and three-stage observers and the Kalman filter, as well observer- and Kalman filter-driven controllers (hence, it can be also extended to stochastic systems) including their two- and three-time-scale counterparts.
(f) The design is independent for each local subsystem so that it provides flexibility for the development of partial full-state feedback (for the subsystem for which all state variables are available for feedback) and partial output feedback (for subsystem for which only the output signal is available for feedback).
(g) Robustness and reliability can be facilitated by using the two- and three-stage designs, as well as the feedback control-loop security can be improved, which appears to be very important these days, especially for cyber physical systems.
(h) The design of desired local controllers, observers, and filters might be feasible even in the cases when the global system is not controllable (observable), but the local subsystems are controllable (observable), in which cases those controllers, observers, and filters could have been designed for particular subsystems of the linear dynamic system.

The statement made in item (h) can be demonstrated by using the following simple example.

Example 1.1 Consider the linear continuous-time system

$$\frac{dx(t)}{dt} = \begin{bmatrix} \frac{dx_1(t)}{dt} \\ \frac{dx_2(t)}{dt} \end{bmatrix} = \begin{bmatrix} a_1 & 0 \\ 0 & a_2 \end{bmatrix} \begin{bmatrix} x_1(t) \\ x_2(t) \end{bmatrix} + \begin{bmatrix} b_1 \\ b_2 \end{bmatrix} u(t), \quad b_1 \neq 0, \quad b_2 \neq 0$$

Finding the controllability matrix, that is,

$$\mathrm{C}(A, b) = [B \ AB] = \begin{bmatrix} b_1 & a_1 b_1 \\ b_2 & a_2 b_2 \end{bmatrix}$$

it can be easily seen that this system is not controllable for $a_1 = a_2$. However, the local subsystems possess local controllability for $b_1 \neq 0, \ b_2 \neq 0$ so that the local linear-quadratic optimal controllers can be designed or the local closed-loop system eigenvalues can be placed in the desired location. A similar example may be constructed for local observability.

The two-, three-, and four-stage feedback designs are applicable to almost all classes of discrete- and continuous-time linear time-invariant systems as well as to linearized time-invariant systems. The feasibility conditions will be spelled out in the corresponding chapters when we present these multistage feedback design techniques.

In this book, we will discuss the possibilities of extending the two-, three-, and four-stage feedback design techniques to general multistage feedback designs of large-scale (complex) systems such that good features (a)–(g) of the two- and three-stage feedback design are preserved. Some of the above features, specially feature (g), appear to be extremely important these days for large-scale systems such as smart power grids, Internet, communication networks, and networks in systems biology and chemistry. In Fig. 1.2, we present a symbolic schematics of the multistage feedback controller design of linear systems with different controllers designed independently for different parts (subsystems) of the system.

The linear time-invariant systems to be subjected to two-, three-, and four-stage and in general multi-stage feedback designs should be first appropriately partitioned and their subsystems identified. The partitioning can be done using several criteria:

(a) Based on the physical nature of the subsystem parts (system natural decomposition)
(b) According to the conditions that must be satisfied such that the partitioned system is feasible for the multistage feedback design
(c) Based on mathematical conditions that must be satisfied to solve the corresponding design equations
(d) Control needs (which parts of the system should be independently controlled via local feedback controllers)
(e) Grouping the state space variables such that the subsystems satisfy control-oriented assumptions (conditions) needed for the design of local controllers, observers, or filters such as controllability (stabilizability) and/or observability (detectability)

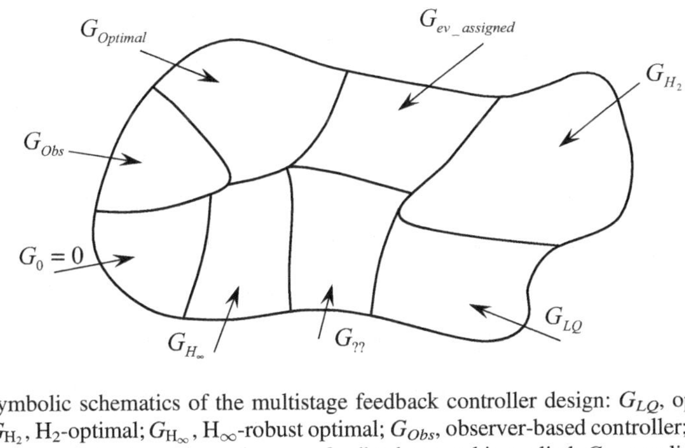

Fig. 1.2 Symbolic schematics of the multistage feedback controller design: G_{LQ}, optimal linear-quadratic; G_{H_2}, H_2-optimal; G_{H_∞}, H_∞-robust optimal; G_{Obs}, observer-based controller; $G_{ev\_assigned}$, eigenvalue assignment; $G_0 = 0$, no subsystem feedback control is applied; $G_{??}$, any linear feedback controller

1.1 General Remarks

The results to be presented in the first four chapters on the two- and three-stage feedback controller designs for continuous- and discrete-time linear dynamic systems represent already the mature controller system design techniques published by the authors in the journals and presented at the conferences. Presently, the authors are finalizing the work on three-stage feedback controller design for discrete-time systems and the four-stage feedback controller design for continuous-time systems. These results are also specialized to corresponding three- and four-time-scale linear systems with applications to proton-exchange membrane fuel cells. From our experience, *the more design stages are involved, the more efficient the controllers become, but conditions for the applicability of these controllers and solutions of corresponding nonlinear algebraic equations become more and more difficult*. The main ideas and complete derivations of discrete-time three-stage feedback design and four-stage continuous-time feedback controller design are presented, respectively, in Chaps. 5 and 6, but these areas still remain open for research, especially the applications to the three-time-scale discrete-time linear systems and to the four-time-scale continuous-time linear dynamic time-invariant systems.

1.2 Applications to Other Classes of Dynamic Systems

The results presented in this research monograph are for linear continuous- and discrete-time time-invariant dynamic systems. The results obtained can be directly extended to the linearized models of nonlinear dynamic systems. The authors believe that this study can be also extended to linearized models of distributed parameter systems described in the modal coordinates by infinite sets of second-order ordinary differential equations, models of flexible space structures, and second-order nonclassically damped linear mechanical systems. Linearized models of distributed parameter systems (infinite dimensional systems described by partial differential equations) can be represented in the modal coordinates by infinite sets of second-order ordinary differential equations (Meirovitch and Baruh 1983; Baruh and Choe 1990). Flexible structures, especially large space flexible structures, can be modeled by an infinite series of pure oscillators (having eigenvalues on the imaginary axis) and lightly damped oscillators (having eigenvalues in the stable half complex plane very close to the imaginary axis) (Gawronski and Juang 1990; Gawronski 1994, 1998). These techniques will facilitate conditions for applications of two-, three-, and four-stage feedback controller design techniques to the corresponding classes of dynamic systems.

1.3 Improved System Robustness, Reliability, and Security

Designing individual controllers for each local subsystem will improve the system robustness, reliability, and security, which are particularly important these days for large (complex)-scale systems such as power grids, water supply networks, Internet, communication networks, biological networks, chemical networks, and in general cyber physical systems. Note that in the multistage feedback design techniques, the total number of the feedback loops remains the same as for the original full-state feedback systems, but the feedback loops are grouped according to the subsystems and feedback algorithms used, hence naturally less prompt to global failures. If one group of the feedback loops fails, the remaining groups will be working properly, hopefully still providing a satisfactory system performance. We expect that, in the future, the multistage feedback design can be successfully extended to control of cyber physical systems and corresponding linear-quadratic dynamic games (Pasqualetti et al. 2015; Zhu and Basar 2015). For example, some feedback gains can be specifically designed to reject external malicious attacks on particular subsystems.

1.4 Book Organization

After the introductory chapter, in Chap. 2, the *continuous-time* two-stage feedback controller design technique of linear time-invariant dynamic systems is presented in detail. All design algebraic and differential equations, expressions for the newly derived matrix formulas needed for the design, as well as the conditions under which the two-stage design is applicable are presented. The results obtained are specialized to the two-time scale singularly perturbed linear systems, emphasizing the design simplification for this class of linear time-invariant systems. To that end, the presentation follows the fundamental results of Radisavljevic-Gajic and Rose (2014). The design is first demonstrated on a tenth-order hydrogen gas reformer (gas processing system) that produces hydrogen from hydrogen-rich fuels. The produced hydrogen is needed for the PEM fuel cell operation. The eigenvalue assignment controller design for both slow and fast subsystems is presented. In addition, a hybrid linear-quadratic optimal slow and eigenvalue assigned fast controller are designed for this two-time scale system.

In the second part of this chapter, we apply the two-stage design methodology to the eight-order model of a PEM fuel cell following the results of Radisavljevic-Gajic et al. (2015). We present also efficient fixed-point and Newton method type numerical algorithms for solving efficiently the design algebraic equations as systems of linear algebraic equations. At the end of this chapter, we design an observer for the considered fuel cell model to estimate all state variables needed for the full-state feedback controller design (either eigenvalue assigned or linear-quadratic optimal controllers). Like in the case of the hydrogen gas reformer, we independently design for this fuel cell the eigenvalue assignment controllers for both the slow and fast subsystems, as well hybrid optimal slow and eigenvalue assigned fast controller for the overall PEM fuel cell model.

Chapter 3 considers the *discrete-time* two-stage feedback controller design. The presentation follows the results obtained in Radisavljevic-Gajic (2015a, b). Like in the continuous time, all design algebraic and differential equations, expressions for the newly derived matrix formulas needed for the design, and the conditions under which the two-stage design is applicable are derived and established. In addition, the obtained results are specialized and simplified to the two-time scale discrete time-invariant systems. Both the slow time scale formulation and the fast time scale formulation of this class of systems are considered. The design efficiency was demonstrated using examples of an electric power system (slow time scale formulation) and a steam power system (fast time scale formulation).

Chapter 4 presents the *continuous-time* three-stage feedback controller design. We have first indicated specifics of the three-stage feedback controller design and differences comparing to the two-stage feedback controller design. We have presented all design formulas, derived corresponding algebraic equations, and established conditions under which such a design is possible. Like in Chaps. 2 and 3, we have specialized and simplified the design to the three-time-scale singularly perturbed systems and shown how to solve all required nonlinear equations as systems of linear algebraic equations using the fixed-point iterations. The efficiency of the presented three-stage three-time feedback controller design is demonstrated on the example of an eight-order PEM fuel cell model. Several controller types were designed: (1) eigenvalue assignment controllers for all three subsystems; (2) optimal linear-quadratic controller which is designed only for the slow subsystem with no controllers used for the fast and very fast subsystems (partial optimization); and (3) optimal linear-quadratic controllers which are designed independently for each subsystem (slow, fast, and very fast) using the subsystem data only. The presentation of this chapter follows closely the recent papers of the authors Radisavljevic-Gajic and Milanovic (2016) and Radisavljevic-Gajic et al. (2017).

Chapter 5 parallels the continuous-time derivations of Chap. 4 and completely derives the three-stage feedback controllers for the general class of linear time-invariant *discrete-time* systems. The chapter also discusses the main ideas and potential derivations of the discrete-time three-stage three-time scale feedback controller design. Presently, the problem formulation for this class of systems is not settled down, and there are at least four different formulations that potentially can be used for the three-stage three-time-scale feedback controller design of linear discrete-time systems. Hence, this topic remains an interesting open area for future research.

Chapter 6 presents the complete derivations for the *continuous-time* four-stage feedback controller design of linear time-invariant systems. Even though the results reported in this chapter are an extension of the results reported in Chap. 4 for the continuous-time three-stage feedback controller design, these derivations are different than those in Chap. 4. The derivations here are much more involved, and the obtained algebraic equations needed for the design are much more complex. No obvious method exits at this time how to solve these algebraic equations in general. Hence, they will be the subject of the future research. We hope that the algebraic equations derived will be considerably simplified in the case of four-time scale singularly perturbed systems, where the small singular perturbation parameters present in these algebraic equations might provide ideas how to solve these equations either using the fixed-point iterations or the Newton method. This formulates another open research problem in this control systems area.

Chapter 7 discusses modeling, system analysis, and control issues with potential limitations and constraints for proton-exchange membrane fuel cell mathematical models. We start with the linear third-order model of El-Sharkh et al. (2004) and the bilinear third-order model of Gemmen (2003) and Chiu et al. (2004) that consider three fundamental fuel cell state space variables: hydrogen pressure, oxygen pressure, and the cathode side water vapor pressure. We show that these third-order models have some fundamental controllability problems. The presentations of the third-order fuel cell mathematical models follow closely the author's papers Radisavljevic (2011) and Radisavljevic-Gajic and Graham (2017). In addition, in the first part of this chapter, we discuss also the fifth-order nonlinear model developed by Na and Gou (2008) and considered in detail in Gou et al. (2010).

In the second part of this chapter, we discuss also the nonlinear fifth-order fuel cell model derived in Milanovic et al. (2017) for the Greenlight Innovation G60 Testing station fuel cell used in the corresponding Villanova University laboratory. State space variables of this model are mass of oxygen in cathode, mass of nitrogen in cathode, mass of hydrogen in anode, mass of water vapor in anode, and mass of water vapor in cathode. The model shows very good agreement with the experimental results. At the end of this chapter, an eight-order model of a PEM fuel cell used in automotive applications is presented based on the results reported in Pukrushpan et al. (2004a, b). In addition to mass of oxygen in cathode, mass of nitrogen in cathode, mass of hydrogen in anode, mass of water vapor in anode, and mass of water vapor in cathode, the model of Pukrushpan et al. (2004a, b) as the state variables have the dynamics of the gas in intake (supply) and outtake (return) manifolds due to the fuel cell particular application for electric cars. This model can be studied in two, three, and eventually four time scales, and it is considered in several case studies in this monograph.

Chapter 8 presents some fundamental control strategies for a hydrogen natural gas processing system (known also as the hydrogen gas reformer) that provides pure hydrogen from the natural gas via simple physical processes with efficiency of roughly 50%. That hydrogen is then used in PEM fuel cells. The chapter is based on the author's research paper Radisavljevic-Gajic and Rose (2015). A reduced-order observer is designed first to estimate the state variables of the tenth-order hydrogen gas reformer. The controller proposed has two feedback controllers and one feed-forward controller. The feedback controllers are an integral controller that copes with constant disturbances and the linear-quadratic optimal controller. The feed-forward controller cancels the disturbance caused by the fuel cell current. The results are derived via a rigorous dynamic linear-quadratic optimization. The simulation results show that the controller designed performed very well rejecting a large disturbance very quickly. Even more, the proposed controller outperforms the corresponding full-order observer-based controller used for the same hydrogen processing system.

Chapter 9 discusses ideas and formulates research problems for extensions of the presented methodology to the general multistage feedback controller design for linear dynamic systems composed of N subsystems, including linear dynamic systems that operate in N time scales. N-stage and N-time scale linear feedback controller design is the final research goal. This research monograph is the first step in that direction, and we believe that it will take a couple of years before the general multistage feedback controller design problem is completely solved. Presently, it is an interesting and challenging control engineering research area.

1.5 Notes

Chapters 2, 3, 4, 7, and 8 use in parts the material from our previous published journal and conference papers. The remaining four chapters are our original contributions. Permissions for the use of such material in this research monograph were granted to us by the American Society of Mechanical Engineers (ASME) for three journal and five conference papers and by Elsevier for two journal papers. Acknowledgments of granted permissions for each particular paper are done at the end of each chapter clearly identifying the particular book sections, papers, and the permission granting publisher.

Chapter 2
Continuous-Time Two-Stage Feedback Controller Design

In this chapter, we first present a general algorithm for a two-stage feedback controller design for linear continuous-time, time-invariant, dynamic systems following the results of Radisavljevic-Gajic and Rose (2014), Sect. 2.1. The proposed design significantly reduces the computational requirements and provides flexibility of designing different types of controllers for different dynamic parts of the system – subsystems that form the given system.

The newly proposed design is further simplified and specialized for linear dynamic systems with slow and fast modes (singularly perturbed linear systems) in Sect. 2.2. The corresponding algorithm is applied efficiently for design of feedback controllers for a hydrogen gas reformer that produces hydrogen (from hydrogen rich fuels like natural gas or methanol) to be used in fuel cells, Sect. 2.3. In Sects. 2.4 and 2.5, we demonstrate the use of the presented algorithm for design of feedback controllers for a proton-exchange membrane fuel cell, including the design of observer-based controllers.

2.1 Two-Stage Design of Linear Feedback Controllers

Design of linear state feedback controllers plays an important role in engineering applications (Ogata 1995; Sinha 2007; Chen 2012). In this section, a two-stage design algorithm developed by Radisavljevic-Gajic and Rose (2014) is presented.

Consider a linear time-invariant dynamic system represented in its partitioned form by

V. Radisavljević-Gajić et al., *Multi-Stage and Multi-Time Scale Feedback Control of Linear Systems with Applications to Fuel Cells*, Mechanical Engineering Series,
https://doi.org/10.1007/978-3-030-10389-7_2

$$\frac{dx(t)}{dt} = \begin{bmatrix} \frac{dx_1(t)}{dt} \\ \frac{dx_2(t)}{dt} \end{bmatrix} = \begin{bmatrix} A_{11} & A_{12} \\ A_{21} & A_{22} \end{bmatrix} \begin{bmatrix} x_1(t) \\ x_2(t) \end{bmatrix} + \begin{bmatrix} B_{11} \\ B_{22} \end{bmatrix} u(t) = Ax(t) + Bu(t) \quad (2.1)$$

where $x(t) \in \mathrm{R}^n$, $x_1(t) \in \mathrm{R}^{n_1}$, $x_2(t) \in \mathrm{R}^{n_2}$, and $n = n_1 + n_2$ represent state space variables, $u(t) \in \mathrm{R}^m$ is the system control input vector, and A_{ij} and B_{ii} and $i,j = 1,2,$ are constant matrices of appropriate dimensions. Matrices A_{11} and A_{22} define subsystems of dimensions n_1 and n_2, respectively, corresponding to the state variables $x_1(t)$ and $x_2(t)$. Matrices A_{12} and A_{21} define couplings between the subsystems.

In the follow-up of the section, it will be shown how to simplify design of independent controllers that operate independently on different subsystems (represented by A_{11} and A_{22}) despite strong coupling among subsystems (represented by A_{12} and A_{21}) and both subsystems having a common input $u(t)$.

In the following, we will show how the subsystem feedback gains can be found independently and then implemented as a compounded full-state feedback gain; see Fig. 1.1.

The new two-stage linear feedback algorithm is based on finding appropriate forms of two similarity transformations (Chen 2012), applied successively to system defined in (2.1) to achieve a system block triangular form. At the end of the design, the linear feedback gain is obtained in the original system coordinates defined in (2.1).

The first results about the two-stage design of linear feedback controllers in both continuous- and discrete-time domains appeared in the classic works of Philips (1980a, b, 1983). The work of Phillips was simplified, and an algorithm that reduces the computational requirements was presented in Radisavljevic-Gajic and Rose (2014). The two-stage linear feedback controller design of Radisavljevic-Gajic and Rose (2014) is presented in five steps given below.

Stage 1

Step 1: Apply the following change of variables to (2.1)

$$\eta(t) = Lx_1(t) + x_2(t) \quad (2.2)$$

where L satisfies a nonsymmetric, nonsquare, Riccati algebraic equation defined by

$$LA_{11} - A_{22}L - LA_{12}L + A_{21} = 0 \quad (2.3)$$

The nonsymmetric algebraic equation (2.3) appears in some areas of systems and control theory. It has been extensively studied by control systems and applied mathematics researchers; see, for example, Medanic (1982), Gao and Bai (2010), and references therein. It can be solved in general using the eigenvector method, Medanic (1982).

The original and new state variables are related via the following linear transformation

$$\begin{bmatrix} x_1(t) \\ \eta(t) \end{bmatrix} = T_{1new} \begin{bmatrix} x_1(t) \\ x_2(t) \end{bmatrix} = \begin{bmatrix} I & 0 \\ L & I \end{bmatrix} \begin{bmatrix} x_1(t) \\ x_2(t) \end{bmatrix} \tag{2.4}$$

and its inverse

$$\begin{bmatrix} x_1(t) \\ x_2(t) \end{bmatrix} = T_{1new}^{-1} \begin{bmatrix} x_1(t) \\ \eta(t) \end{bmatrix} = \begin{bmatrix} I & 0 \\ -L & I \end{bmatrix} \begin{bmatrix} x_1(t) \\ \eta(t) \end{bmatrix} \tag{2.5}$$

The application of transformation (2.2) to (2.1) produces

$$\begin{aligned} \begin{bmatrix} \frac{dx_1(t)}{dt} \\ \frac{d\eta(t)}{dt} \end{bmatrix} &= \begin{bmatrix} A_{11} - A_{12}L & A_{12} \\ 0 & A_{22} + LA_{12} \end{bmatrix} \begin{bmatrix} x_1(t) \\ \eta(t) \end{bmatrix} + \begin{bmatrix} B_{11} \\ B_{22} + LB_{11} \end{bmatrix} u(t) \\ &= \begin{bmatrix} A_1 & A_{12} \\ 0 & A_2 \end{bmatrix} \begin{bmatrix} x_1(t) \\ \eta(t) \end{bmatrix} + \begin{bmatrix} B_{11} \\ B_2 \end{bmatrix} u(t) \end{aligned} \tag{2.6}$$

where

$$A_1 = A_{11} - A_{12}L, \quad A_2 = A_{22} + LA_{12}, \quad B_2 = B_{22} + LB_{11} \tag{2.7}$$

Step 2: Use feedback control

$$u(t) = -G_2\eta(t) + v(t)$$

to set up the closed-loop eigenvalues of the η-subsystem, which leads to the following linear system

$$\begin{bmatrix} \frac{dx_1(t)}{dt} \\ \frac{d\eta(t)}{dt} \end{bmatrix} = \begin{bmatrix} A_1 & A_{12} - B_{11}G_2 \\ 0 & A_2 - B_2G_2 \end{bmatrix} \begin{bmatrix} x_1(t) \\ \eta(t) \end{bmatrix} + \begin{bmatrix} B_{11} \\ B_2 \end{bmatrix} v(t) \tag{2.8}$$

Stage 2

Step 3: Apply another change of state variables as

$$\xi(t) = x_1(t) - P\eta(t) \tag{2.9}$$

where P satisfies the linear Sylvester algebraic equation (Chen 2012)

$$A_1P - P(A_2 - B_2G_2) + A_{12} - B_{11}G_2 = 0 \tag{2.10}$$

with matrices A_1, A_2, B_2 previously defined in (2.7). This defines another similarity transformation

$$\begin{bmatrix} \xi(t) \\ \eta(t) \end{bmatrix} = T_{2new} \begin{bmatrix} x_1(t) \\ \eta(t) \end{bmatrix} = \begin{bmatrix} I & -P \\ 0 & I \end{bmatrix} \begin{bmatrix} x_1(t) \\ \eta(t) \end{bmatrix} \tag{2.11}$$

whose inverse is

$$\begin{bmatrix} x_1(t) \\ \eta(t) \end{bmatrix} = T_{2new}^{-1} \begin{bmatrix} \xi(t) \\ \eta(t) \end{bmatrix} = \begin{bmatrix} I & P \\ 0 & I \end{bmatrix} \begin{bmatrix} x_1(t) \\ \eta(t) \end{bmatrix} \tag{2.12}$$

which leads to

$$\begin{bmatrix} \dfrac{d\xi(t)}{dt} \\ \dfrac{d\eta(t)}{dt} \end{bmatrix} = \begin{bmatrix} A_1 & 0 \\ 0 & A_2 - B_2G_2 \end{bmatrix} \begin{bmatrix} \xi(t) \\ \eta(t) \end{bmatrix} + \begin{bmatrix} B_{11} - PB_2 \\ B_2 \end{bmatrix} v(t) \tag{2.13}$$

Step 4: Use state feedback

$$v(t) = -G_1\xi(t)$$

for the ξ-subsystem, which produces

$$\begin{bmatrix} \dfrac{d\xi(t)}{dt} \\ \dfrac{d\eta(t)}{dt} \end{bmatrix} = \begin{bmatrix} A_1 - (B_{11} - PB_2)G_1 & 0 \\ -B_2G_1 & A_2 - B_2G_2 \end{bmatrix} \begin{bmatrix} \xi(t) \\ \eta(t) \end{bmatrix} \tag{2.14}$$

Since (2.14) is a lower triangular matrix, its eigenvalues are the union of the eigenvalues of $A_2 - B_2G_2$ and $A_1 - (B_{11} - PB_2)G_1$, both set up at the subsystem levels.

Step 5: The feedback gains in the original coordinates are obtained using the similarity transformations as follows

$$\begin{aligned}
u(x(t)) &= -[0 \;\; G_2]\begin{bmatrix} x_1(t) \\ \eta(t) \end{bmatrix} - [G_1 \;\; 0]\begin{bmatrix} \xi(t) \\ \eta(t) \end{bmatrix} \\
&= -[0 \;\; G_2]\begin{bmatrix} x_1(t) \\ \eta(t) \end{bmatrix} - [G_1 \;\; 0]T_{2new}\begin{bmatrix} x_1(t) \\ \eta(t) \end{bmatrix} \\
&= -\{[0 \;\; G_2] + [G_1 \;\; 0]T_{2new}\}T_{1new}\begin{bmatrix} x_1(t) \\ x_2(t) \end{bmatrix} \\
&= -\left\{[0 \;\; G_2] + [G_1 \;\; 0]\begin{bmatrix} I & -P \\ 0 & I \end{bmatrix}\right\}\begin{bmatrix} I & 0 \\ L & I \end{bmatrix}\begin{bmatrix} x_1(t) \\ x_2(t) \end{bmatrix} \\
&= -[G_1 \;\; G_2 - G_1 P]\begin{bmatrix} I & 0 \\ L & I \end{bmatrix}\begin{bmatrix} x_1(t) \\ x_2(t) \end{bmatrix} \\
&= -[G_1 + (G_2 - G_1 P)L \;\;\; G_2 - G_1 P]\begin{bmatrix} x_1(t) \\ x_2(t) \end{bmatrix} \\
&= -G_{1eq}x_1(t) - G_{2eq}x_2(t) = -G_{eq}x(t) = -[G_{1eq} \;\;\; G_{2eq}]x(t)
\end{aligned} \tag{2.15}$$

with the equivalent feedback gains from the original state space variables equal to

$$G_{1eq} = G_1 + (G_2 - G_1 P)L, \quad G_{2eq} = G_2 - G_1 P \tag{2.16}$$

The proposed designs will require solutions of systems of nonsquare nonsymmetric algebraic equations. Note that the nonsymmetric, nonsquare, algebraic equation has many real solutions (Medanic 1982) and *any real solution in general will serve the purpose of the considered two-stage design* of linear feedback controllers. The proposed technique in general is not unique (due to nonuniqueness of L) which is an advantage so that the presented design can provide several decomposition pairs of subsystems, and for its implementability, it is sufficient that only one of the subsystem pairs satisfies the design conditions (assumptions). A solution of the nonsymmetric, nonsquare algebraic equation defined in (2.3) can be obtained via the eigenvector method from the generalized eigenvectors of the corresponding $n \times n$ matrix H; see, for example, Medanic (1982) and Bingulac and Van Landingham (1993).

$$H = \begin{bmatrix} A_{11} & A_{12} \\ A_{21} & A_{22} \end{bmatrix}^{n \times n} \tag{2.17}$$

Using, for example, the algorithm of Bingulac and Van Landingham (1993), it is required that a matrix V is formed from corresponding real eigenvectors of H, and, for all complex-conjugate eigenvectors of H, we put in matrix V both its real and imaginary parts and discard their complex-conjugate pairs. Partitioning the matrix V as

$$V^{n\times n} = \begin{bmatrix} V_1^{n\times n_1} & V_2^{n\times n_2} \end{bmatrix} = \begin{bmatrix} V_{11}^{n_1\times n_1} & V_{21}^{n_1\times n_2} \\ V_{12}^{n_2\times n_1} & V_{22}^{n_2\times n_2} \end{bmatrix} \tag{2.18}$$

any solution for L can be obtained using the formula (Medanic 1982; Bingulac and Van Landingham 1993)

$$L = V_{12}V_{11}^{-1} \tag{2.19}$$

Hence, any collection of n_1 eigenvectors of matrix H that provides invertible matrix V_{11} will serve the purpose of the presented two-stage linear feedback design. Since there are many permutations of the eigenvectors of H, in general, there are many feasible solutions, and hence there are no problems with the existence of a solution of equation (2.3).

2.2 Two-Stage Feedback Design for Systems with Slow and Fast Modes

Linear systems with slow and fast modes (singularly perturbed systems (Kokotovic et al. 1999; Naidu and Calise 2001, also known as multi-time scale systems)) are particularly well suited for the considered two-stage feedback design. For this class of systems, in general, numerical ill-conditioning appears if one attempts to design linear feedback controller using the entire (full-order) system. Singularly perturbed systems have numerous applications in engineering and sciences (Kokotovic et al. 1999; Naidu and Calise 2001) and play an important role in mechanical and aerospace engineering (Hsiao et al. 2001; Naidu and Calise 2001; Chen et al. 2002; Shapira and Ben-Asher 2004; Demetriou and Kazantzis 2005; Wang and Ghorbel 2006; Amjadifard et al. 2011; Kuehn 2015).

In this section, further design simplifications will be achieved by specializing the proposed design from Sect. 2.1 to singularly perturbed linear systems so that only solutions of linear *algebraic equations* will be required. The digital implementation of the corresponding controllers will allow different sampling rates to be used for the slow (large sampling rate) and the fast (small sampling rates) controllers. Otherwise, without the two-stage design, the whole system digital controller will require the small sampling rate. Moreover, for the class of linear dynamic systems, the corresponding two-stage design algebraic equations have unique solutions for sufficiently small values of the singular perturbation parameter.

The corresponding time-invariant linear continuous-time singularly perturbed system is defined by

$$\begin{bmatrix} \dfrac{dx_1(t)}{dt} \\ \varepsilon\dfrac{dx_2(t)}{dt} \end{bmatrix} = \begin{bmatrix} A_{11} & A_{12} \\ A_{21} & A_{22} \end{bmatrix} \begin{bmatrix} x_1(t) \\ x_2(t) \end{bmatrix} + \begin{bmatrix} B_{11} \\ B_{22} \end{bmatrix} u(t)$$
$$y(t) = C_1 x_1(t) + C_2 x_2(t) \tag{2.20}$$

where ε is a small positive singular perturbation parameter that indicates separation of state space variables into slow ones, $x_1(t)$, and the fast state variables $x_2(t)$. The dimensions of state variables, control input, and constant matrices are defined in Sect. 2.1. It is a standard assumption in the theory of singularly perturbed systems that matrix A_{22} is nonsingular (Kokotovic et al. 1999; Naidu and Calise 2001). Hence, the following assumption is imposed in the follow-up of this chapter.

Assumption 2.1 Fast subsystem matrix A_{22} is nonsingular.

In the case of singularly perturbed systems, the proposed simplified two-stage feedback design of Radisavljevic-Gajic and Rose (2014) has the following steps grouped in two stages.

Stage 1

Step 1: Solve the algebraic Riccati-type equation (2.3), which in the case of singularly perturbed systems due to presence of the small positive singular perturbation parameter ε has the form

$$\varepsilon L A_{11} - A_{22} L - \varepsilon L A_{12} L + A_{21} = 0 \tag{2.21}$$

A unique solution of (2.21) exists for sufficiently small values of ε under Assumption 2.1. Under that assumption, (2.21) can be efficiently solved by performing fixed-point iterations on a system of *linear algebraic equations* as follows

$$A_{22} L^{(i+1)} = A_{21} + \varepsilon L^{(i)} A_{11} - \varepsilon L^{(i)} A_{12} L^{(i)}, \quad L^{(0)} = A_{22}^{-1} A_{21}, \quad i = 1,2,..,k \tag{2.22}$$

It can be shown that this algorithm has the rate of convergence of $O(\varepsilon)$, meaning that after i iterations the accuracy of $O(\varepsilon^i)$ is achieved, with $O(\varepsilon^i)$ defined as $O(\varepsilon^i) < c\varepsilon^i$, where c is a bounded constant and i is a real number. Moreover, if ε is not sufficiently small, the eigenvector method (Medanic 1982) can be used to solve (2.21).

Apply the change of variables to (2.20)

$$x_f(t) = L x_1(t) + x_2(t) \tag{2.23}$$

which leads to the following upper block triangular system

$$\begin{bmatrix} \dfrac{dx_1(t)}{dt} \\ \varepsilon\dfrac{dx_f(t)}{dt} \end{bmatrix} = \begin{bmatrix} A_{11} - A_{12}L & A_{12} \\ 0 & A_{22} + \varepsilon LA_{12} \end{bmatrix} \begin{bmatrix} x_1(t) \\ x_f(t) \end{bmatrix} + \begin{bmatrix} B_{11} \\ B_{22} + \varepsilon LB_{11} \end{bmatrix} u(t)$$
$$= \begin{bmatrix} A_s & A_{12} \\ 0 & A_f \end{bmatrix} \begin{bmatrix} x_1(t) \\ x_f(t) \end{bmatrix} + \begin{bmatrix} B_{11} \\ B_f \end{bmatrix} u(t) \tag{2.24}$$

where

$$A_s = A_{11} - A_{12}L, \quad A_f = A_{22} + \varepsilon LA_{12}, \quad B_f = B_{22} + \varepsilon LB_{11} \tag{2.25}$$

The original and new state variables are related via a similarity transformation as follows

$$\begin{bmatrix} \xi(t) \\ \eta(t) \end{bmatrix} = T_{2new} \begin{bmatrix} x_1(t) \\ \eta(t) \end{bmatrix} = \begin{bmatrix} I & -P \\ 0 & I \end{bmatrix} \begin{bmatrix} x_1(t) \\ \eta(t) \end{bmatrix} \tag{2.26}$$

whose inverse is

$$\begin{bmatrix} x_1(t) \\ \eta(t) \end{bmatrix} = T_{2new}^{-1} \begin{bmatrix} \xi(t) \\ \eta(t) \end{bmatrix} = \begin{bmatrix} I & P \\ 0 & I \end{bmatrix} \begin{bmatrix} x_1(t) \\ \eta(t) \end{bmatrix} \tag{2.27}$$

Step 2: Apply feedback control

$$u(t) = -G_f x_f(t) + v(t)$$

to the *fast subsystem*, which leads to

$$\begin{bmatrix} \dfrac{dx_1(t)}{dt} \\ \varepsilon\dfrac{dx_f(t)}{dt} \end{bmatrix} = \begin{bmatrix} A_s & A_{12} - B_{11}G_f \\ 0 & A_f - B_f G_f \end{bmatrix} \begin{bmatrix} x_1(t) \\ x_f(t) \end{bmatrix} + \begin{bmatrix} B_{11} \\ B_f \end{bmatrix} v(t) \tag{2.28}$$

Stage 2

Step 3: Apply another change of state variables as

$$x_s(t) = x_1(t) - \varepsilon P x_f(t) \tag{2.29}$$

where P satisfies the Sylvester algebraic equation

$$\varepsilon A_s P - P\left(A_f - B_f G_f\right) + A_{12} - B_{11} G_f = 0 \tag{2.30}$$

with matrices A_s, A_f, B_f defined in (2.25). Having obtained L from (2.22), the algebraic equation (2.30) can be also solved directly as the Sylvester equation, in which case the unique solution is obtained under the assumption that matrices εA_s and $A_f - B_f G_f$ have no eigenvalues in common (Chen 2012) which is always true for singularly perturbed systems since they have eigenvalues separated into two clusters: the slow ones close to the imaginary axis and the fast ones far from the imaginary axis. Moreover, the slow eigenvalues are multiplied by a small positive parameter ε so that the unique solution of the Sylvester algebraic equation (2.30) always exists since $\lambda(\varepsilon A_s) = \varepsilon\lambda(A_s)$ (Stewart 1973).

Formula (2.29) defines another similarity transformation

$$\begin{bmatrix} x_s(t) \\ x_f(t) \end{bmatrix} = T_{2sp} \begin{bmatrix} x_1(t) \\ x_f(t) \end{bmatrix} = \begin{bmatrix} I & -\varepsilon P \\ 0 & I \end{bmatrix} \begin{bmatrix} x_1(t) \\ x_f(t) \end{bmatrix} \tag{2.31a}$$

$$\begin{bmatrix} x_1(t) \\ x_f(t) \end{bmatrix} = T_{2sp}^{-1} \begin{bmatrix} x_s(t) \\ x_f(t) \end{bmatrix} = \begin{bmatrix} I & \varepsilon P \\ 0 & I \end{bmatrix} \begin{bmatrix} x_s(t) \\ x_f(t) \end{bmatrix} \tag{2.31b}$$

which leads to

$$\begin{bmatrix} \dfrac{dx_s(t)}{dt} \\ \varepsilon\dfrac{dx_f(t)}{dt} \end{bmatrix} = \begin{bmatrix} A_s & 0 \\ 0 & A_f - B_f G_f \end{bmatrix} \begin{bmatrix} x_s(t) \\ x_f(t) \end{bmatrix} + \begin{bmatrix} B_{11} - PB_f \\ B_f \end{bmatrix} v(t) \tag{2.32}$$

Step 4: Use state feedback

$$v(t) = -G_s x_s(t)$$

to set up the closed-loop eigenvalues of the slow subsystem. This produces

$$\begin{bmatrix} \dfrac{dx_s(t)}{dt} \\ \varepsilon\dfrac{dx_f(t)}{dt} \end{bmatrix} = \begin{bmatrix} A_s - \left(B_{11} - PB_f\right)G_s & 0 \\ -B_f G_s & A_f - B_f G_f \end{bmatrix} \begin{bmatrix} x_s(t) \\ x_f(t) \end{bmatrix} \tag{2.33}$$

Since (2.33) is a lower triangular matrix, its eigenvalues are the union of the eigenvalues of $(A_f - B_f G_f)/\varepsilon$ and $A_s - (B_{11} - PB_f)G_s$, both set up at the subsystem levels.

Step 5: The feedback gains in the original system state space coordinates are obtained from

$$
\begin{aligned}
u(x(t)) &= -[0 \quad G_f]\begin{bmatrix} x_1(t) \\ x_f(t) \end{bmatrix} - [G_s \quad 0]\begin{bmatrix} x_s(t) \\ x_f(t) \end{bmatrix} \\
&= -[0 \quad G_f]\begin{bmatrix} x_1(t) \\ x_f(t) \end{bmatrix} - [G_s \quad 0]T_{2sp}\begin{bmatrix} x_1(t) \\ x_f(t) \end{bmatrix} \\
&= -\{[0 \quad G_f] + [G_s \quad 0]T_{2sp}\}T_{1sp}\begin{bmatrix} x_1(t) \\ x_2(t) \end{bmatrix} \\
&= -\left\{[0 \quad G_f] + [G_s \quad 0]\begin{bmatrix} I & -\varepsilon P \\ 0 & I \end{bmatrix}\right\}\begin{bmatrix} I & 0 \\ L & I \end{bmatrix}\begin{bmatrix} x_1(t) \\ x_2(t) \end{bmatrix} \\
&= -[G_s \quad G_f - \varepsilon G_s P]\begin{bmatrix} I & 0 \\ L & I \end{bmatrix}\begin{bmatrix} x_1(t) \\ x_2(t) \end{bmatrix} \\
&= -\left[G_s + (G_f - \varepsilon G_s P)L \quad G_f - \varepsilon G_s P\right]\begin{bmatrix} x_1(t) \\ x_2(t) \end{bmatrix} \\
&= -G_{1eq}x_1(t) - G_{2eq}x_2(t) = -G_{eq}x(t)
\end{aligned}
\tag{2.34}
$$

The equivalent feedback gains from the original state space variables are obtained as

$$
G_{1eq} = G_s + (G_f - \varepsilon G_s P)L = G_s + G_{2eq}L, \quad G_{2eq} = G_f - \varepsilon G_s P \tag{2.35}
$$

In summary, to perform linear feedback design in two stages for singularly perturbed linear systems (including the eigenvalue assignment), it is needed to solve the nonlinear Riccati-type algebraic equation (2.21) and the linear Sylvester algebraic equation (2.30). Nonlinear algebraic equation (2.21) is efficiently solved by performing iterations on systems of linear algebraic equations as demonstrated in (2.22).

2.3 Two-Stage Control of a Hydrogen Gas Reformer Slow-Fast Dynamics

In this section, we demonstrate the use of the new two-stage feedback controller design technique presented in Sect. 2.2 for a hydrogen gas reformer (Pukrushpan et al. 2004a, 2006; Tsourapas et al. 2007; Cipiti et al. 2013) which produces hydrogen from natural gas that is used for PEM fuel cells. Natural gas is a hydrocarbon gas mixture consisting mostly of methane (CH_4) with small amounts of paraffin (saturated hydrocarbon), carbon dioxide, nitrogen, and hydrogen sulfide. To extract hydrogen from natural gas, several simple reactions have to be performed. The hydrogen obtained is pumped to the anode side of the PEM fuel cell (Pukrushpan et al. 2004a).

Modeling and control of the hydrogen gas reformer dynamics has been an important and challenging research area, as discussed in Pukrushpan et al. (2004a, 2006), Tsourapas et al. (2007), and references therein. The mathematical model of the hydrogen gas reformer considered in this section is of a relatively high order (ten), and it has variables that operate in two time scales, slow and fast, which requires additional attention due to potential numerical ill-conditioning (e.g., huge slope at the initial time of the fast state variables). The two-time-scale dynamics comes from different processes that govern dynamics of the hydrogen gas reformer coupled to a PEM fuel cell. Such processes are chemical, electrical, electronic, electrochemical, mechanical, and thermodynamic, and they have different time constants that correspond to the complex dynamics of the hydrogen gas reformer coupled to a PEM fuel cell, both operating in multi-time scales.

2.3.1 *Hydrogen Gas Reformer Operation and Modeling*

Fuel cells utilize chemical reactions with hydrogen gas to produce electricity. However, H_2 gas is not always easily available for a fuel cell system. A solution to this problem is to use a hydrogen gas reformer also known as the fuel processor system (FPS) to purify gas, typically natural gas, into the needed H_2 gas (Pukrushpan et al. 2004a, 2006; Tsourapas et al. 2007; Cipiti et al. 2013). A common process used to extract hydrogen from natural gas in an FPS is partial oxidation. This process uses chemical reactions of natural gas and air to produce a H_2-rich gas product. The four main reactors of the FPS shown in Fig. 2.1 are hydro-desulfurizer (HDS), catalytic partial oxidizer (CPOX), water gas shift (WGS), and preferential oxidizer (PROX). Once the gas travels through all of these reactors, H_2-rich gas will be produced. In Fig. 2.1, HEX stands for a heat exchanger and MIX for a mixer.

Natural gas enters the FPS via a high pressure source, usually a tank or a gas line. The gas is first fed through the hydro-desulfurizer to eliminate any sulfur that could be contained in the gas. This is done because sulfur can poison the water gas shift. The desulfurized gas is then passed to the mixer (MIX), where it blends with air. The air is first brought into the FPS by a blower and then passes through the heat exchanger to reach a necessary temperature.

Once mixed, the gas passes through the catalytic partial oxidizer where a catalyst causes the natural gas to react with the oxygen in the air. Two exothermal reactions

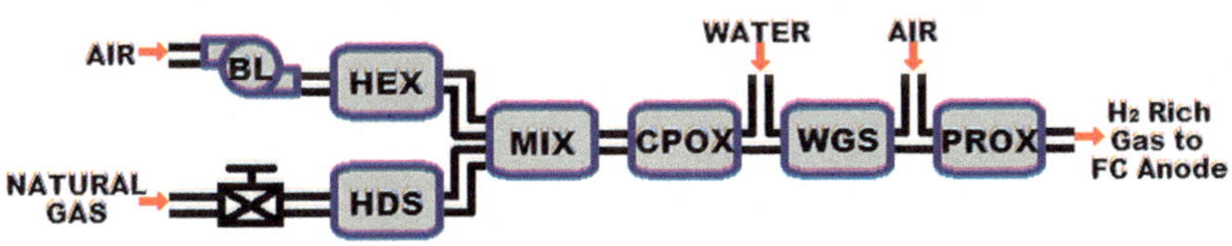

Fig. 2.1 Fuel processing system (FPS)

take place in the CPOX: partial oxidation (POX) and total oxidation (TOX). Partial oxidation produces H_2 gas and carbon monoxide. The total oxidation produces water and carbon dioxide. Even though both reactions generate heat, TOX releases much larger amount of heat ($\Delta H^o_{tox} = -0.8026 \times 10^6 \text{ J/mol}$)

$$\text{(POX)} \quad CH_4 + \frac{1}{2}O_2 \rightarrow CO + 2H_2 \quad \Delta H^o_{pox} = -0.036 \times 10^6 \text{ J/mol}$$

$$\text{(TOX)} \quad CH_4 + 2O_2 \rightarrow CO_2 + 2H_2O \quad \Delta H^o_{tox} = -0.8026 \times 10^6 \text{ J/mol}$$

Since only POX is producing H_2, it is preferable to increase the amount of gas reacting through POX instead of TOX. That is highly dependent on the ratio of O_2 and gas entering CPOX and the CPOX catalyst bed temperature T_{cpox}, Zhu et al. (2001). In addition, very high temperature T_{cpox} can cause CPOX damage, while the low temperature T_{cpox} causes inefficient reactions. Since the reactions in the CPOX, especially H_2 production, are strongly dependent on the CPOX reactor temperature T_{cpox}, it is necessary to have a control mechanism for the temperature.

Even though H_2 is produced by the POX reaction, carbon monoxide is also produced. CO poisons the PEM fuel cell catalyst and therefore needs to be removed. To efficiently solve this problem, the next two reactors are required, the water gas shift and the preferential oxidizer. From the CPOX, the gas mixture flows into the water gas shift, where water is injected into the chamber to react with CO.

$$\text{(WGS)} \quad CO + H_2O \rightarrow CO_2 + H_2$$

The WGS reaction eliminates CO and produces additional H_2. This process does not convert all CO into CO_2, and the mixture is not safe for fuel cell applications. Therefore, the gas mixture is next passed into the PROX, where the remaining CO reacts with the oxygen from the injected air.

$$\text{(PROX)} \quad 2CO + O_2 \rightarrow 2CO_2$$

After leaving the PROX, the gas is rich in H_2 and is safe to be sent to the anode of the PEM fuel cell.

The reformer mathematical model and its controller/observer design techniques are considered in Pukrushpan et al. (2004a, b, 2006) and Tsourapas et al. (2007). The designed observer-based controller helps to regulate temperature of the catalytic partial oxidation process and the anode hydrogen mole fraction at the desired values. The corresponding tenth-order nonlinear mathematical model and its state space variables can be found in Pukrushpan et al. (2004a, 2006) and Tsourapas et al. (2007)

$$\frac{dx(t)}{dt} = f(x(t), u(t), w(t))$$

$$\begin{aligned} x &= [x_1(t) \quad x_2(t) \quad x_3(t) \quad x_4(t) \quad x_5(t) \quad x_6(t) \quad x_7(t) \quad x_8(t) \quad x_9(t) \quad x_{10}(t)]^T \\ &= \left[T_{cpox} \quad p_{H_2} \quad p^{an} \quad p^{hex} \quad \omega^{bl} \quad p^{hds} \quad p^{mix}_{CH_4} \quad p^{mix}_{air} \quad p^{wrox}_{H_2} \quad p^{wrox}\right] \end{aligned} \tag{2.36}$$

The state variables represent the following quantities:

$x_1(t) = T_{cpox}(t)$ – catalyst temperature
$x_2(t) = p^{an}_{H_2}(t)$ – hydrogen pressure in the anode
$x_3(t) = p^{an}(t)$ – anode pressure
$x_4(t) = p^{hex}(t)$ – heat exchanger pressure
$x_5(t) = \omega^{bl}(t)$ – compressor blower angular velocity (rad/s)
$x_6(t) = p^{hds}(t)$ – hydro-desulfurizer pressure
$x_7(t) = p^{mix}_{CH_4}(t)$ – methane (CH_4) pressure in the mixer
$x_8(t) = p^{mix}_{air}(t)$ – air pressure in the mixer
$x_9(t) = p^{wrox}_{H_2}(t)$ – hydrogen pressure in the gas shift converter
$x_{10}(t) = p^{wrox}(t)$ – total pressure in the gas shift converter

The compressor blows air needed for the fuel (natural gas) oxidation. In the model defined in (2.36), $w(t)$ is the disturbance, and it represents the fuel cell stack (connected to the hydrogen reformer) current, $w(t) = I_{st}(t)$, given by $I_{st}(t) = V_{st}(t)/R_L$, where $V_{st}(t)$ is the fuel cell stack voltage and R_L stands for the fuel cell stack load.

The control variables operate the blower's angular velocity and the fuel (natural gas) tank valve, that is,

$$u(t) = \begin{bmatrix} u_{blower}(t) \\ u_{valve}(t) \end{bmatrix} \tag{2.37}$$

The measured output $y(t)$ serves also as the controlled variable $y(t) = z(t)$. It is given by

$$y(t) = z(t) = \left[T_{cpox}(t) \quad y^{an}_{H_2}(t)\right]^T \tag{2.38}$$

where $y^{an}_{H_2}(t)$ is the anode hydrogen mole fraction.

The control objective is to regulate the catalytic partial oxidation temperature and the anode hydrogen mole fraction at the desired values at steady state $T_{cpox} = 972$ K (that corresponds to the ratio of the number of oxygen moles over the number of methane moles equal to 0.6) and $y^{an}_{H_2} = 0.088$ (8.8%) (corresponding to utilization of 80%). In this section, we consider the hydrogen gas reformer linearized mathematical model (Pukrushpan et al. 2004a, b, 2006; Tsourapas et al. 2007). The linearized system model is represented by the following differential equations

$$\begin{gathered}\frac{\delta x(t)}{dt} = A\delta x(t) + B\delta u(t) + \Gamma\delta w(t)\\ \delta y(t) = \delta z(t) = C\delta x(t)\\ y(t) = y_{ss} + \delta y(t), \quad z(t) = z_{ss} + \delta z(t), \quad x(t) = x_{ss} + \delta x(t)\\ u(t) = u_{ss} + \delta u(t), \quad w(t) = w_{ss} + \delta w(t)\end{gathered} \tag{2.39}$$

$\delta x(t)$, $\delta u(t)$, $\delta y(t)$, $\delta z(t)$, $\delta w(t)$ represent variations of the corresponding quantities with respect to their steady-state values. Since the stack current changes as $I_{st} = V_{st}(t)/R_L$, where the load R_L changes in time as a piecewise constant function (Tsourapas et al. 2007), the design of controllers with integral action (Khalil 2002) will be required to cope with the piecewise constant disturbance $\delta w(t) = \delta I_{st}(t)$. For the purpose of this section and to emphasize the newly presented two-stage design of feedback controllers, it is assumed that $\delta w(t) = \delta I_{st}(t) = 0$.

The linearized system matrices obtained by following the linearization procedure of nonlinear systems at nominal (operating) points (Khalil 2002) at the desired steady state can be found in page 147 of the book by Pukrushpan et al. (2004a). It is important to observe that the hydrogen gas reformer model is controllable (Sinha 2007; Chen 2012) so that corresponding linear feedback controllers can assign the closed-loop eigenvalues that can be placed in any desired location in the complex plane.

The state space matrices (corresponding to 50% current load) are given by

$$A = \begin{bmatrix} -0.074 & 0 & 0 & 0 & 0 & 0 & -3.53 & 1.0748 & 0 & 0\\ 0 & -1.468 & -25.3 & 0 & 0 & 0 & 0 & 0 & 2.5582 & 13.911\\ 0 & 0 & -156 & 0 & 0 & 0 & 0 & 0 & 0 & 33.586\\ 0 & 0 & 0 & 124.5 & 212.63 & 0 & 112.69 & 112.69 & 0 & 0\\ 0 & 0 & 0 & 0 & -3.3333 & 0 & 0 & 0 & 0 & 0\\ 0 & 0 & 0 & 0 & 0 & -32.43 & 32.304 & 32.304 & 0 & 0\\ 0 & 0 & 0 & 0 & 0 & 331.8 & -344 & -341 & 0 & 9.9042\\ 0 & 0 & 0 & 221.97 & 0 & 0 & -253.2 & -254.9 & 0 & 32.526\\ 0 & 0 & 2.0354 & 0 & 0 & 0 & 1.8309 & 1.214 & -0.358 & -3.304\\ 0.0188 & 0 & 8.1642 & 0 & 0 & 0 & 5.6043 & 5.3994 & 0 & -13.61 \end{bmatrix}$$

$$B = \begin{bmatrix} 0 & 0 & 0 & 0 & 0.12 & 0 & 0 & 0 & 0 & 0\\ 0 & 0 & 0 & 0 & 0 & 0.1834 & 0 & 0 & 0 & 0 \end{bmatrix}^T$$

$$\Gamma = \begin{bmatrix} 0 & -0.328 & -0.024 & 0 & 0.0265 & 0.0504 & 0 & 0 & 0 & 0 \end{bmatrix}^T$$

$$C = \begin{bmatrix} 1 & 0 & 0 & 0 & 0 & 0 & 0 & 0 & 0 & 0\\ 0 & 0.994 & -0.088 & 0 & 0 & 0 & 0 & 0 & 0 & 0 \end{bmatrix}$$

The eigenvalue assignment is performed in two stages by calculating independently slow and fast subsystem feedback gains using the reduced-order dynamic model system matrices and corresponding reduced-order matrices of the hydrogen gas reformer. The open-loop eigenvalues of this system are separated into three slow eigenvalues located very close to the imaginary at $-0.0862, -0.358, -1.468$ and seven fast eigenvalues located further from the imaginary axis at

$$-2.771 \pm j0.5473, -3.333, -12.169, -89.137, -157.9, -660.68$$

It can be shown after many attempts that there exists a similarity transformation, $\bar{x}(t) = Tx(t)$, $\bar{A} = TAT^{-1}$, $B_{11} = TB$, which puts the hydrogen gas reformer model into the standard singularly perturbed form defined by (2.20) with $n_1 = 3$ and $n_2 = 7$. For the small singular perturbation parameter taken as $\varepsilon = 0.52 = 1.468/2.771$, the corresponding matrices according to the notation used in (2.20) are given by

$$A_{11} = \begin{bmatrix} -0.4949 & 5.5762 & -5.386 \\ -0.0664 & 0.6406 & -0.8501 \\ -0.1509 & 2.0413 & -2.0545 \end{bmatrix}$$

$$A_{12} = \begin{bmatrix} -0.2371 & 0.8823 & 56.287 & 2.3071 & 0.2255 & 5.7138 & 5.3861 \\ 1.987 & 0.1184 & 7.5552 & 0.9791 & 1.9266 & 1.2804 & 0.8501 \\ -0.0825 & 0.2690 & 17.162 & 2.8748 & 0.2810 & -0.2569 & 2.0545 \end{bmatrix}$$

$$A_{21} = \begin{bmatrix} 0 & 0 & 0 \\ 0 & 0 & 0 \\ 0 & 0 & 0 \\ 0 & 0 & 0 \\ 0 & 0 & 0 \\ 0 & 0 & 0 \\ 0.0032 & 0 & -0.0087 \end{bmatrix}$$

$$A_{22} = \begin{bmatrix} -156 & 0 & 0 & 0 & 0 & 0 & 33.586 \\ 0 & -124.5 & 212.63 & 0 & 112.69 & 112.69 & 0 \\ 0 & 0 & -3.333 & 0 & 0 & 0 & 0 \\ 0 & 0 & 0 & -32.43 & 32.304 & 32.304 & 0 \\ 0 & 0 & 0 & 331.8 & -344 & -341 & 9.9042 \\ 0 & 221.97 & 0 & 0 & -252.2 & -254.9 & 32.526 \\ 8.1647 & -0.0057 & -0.3629 & 0.2139 & 5.6252 & 5.3962 & -13.601 \end{bmatrix}$$

The solution of the L-equation (2.21) is obtained by using the fixed-point algorithm (2.22), run for $k = 21$ iterations to obtain the accuracy of $E^{(k)} = O(10^{-15})$, where the accuracy is defined by a norm of the following expression

$$E^{(k)} = \varepsilon L^{(k)} A_{11} - A_{22} L^{(k)} - \varepsilon L^{(k)} A_{12} L^{(k)} + A_{21}$$

Having obtained $L \approx L^{(21)}$, matrices A_s, A_f, B_f are to be calculated from (2.25)

$$L=\begin{bmatrix} -0.0001 & 0 & 0.0002 \\ -0.0002 & 0 & 0.0006 \\ 0 & 0 & 0 \\ -0.0003 & 0 & 0.0007 \\ -0.0005 & 0 & 0.0013 \\ 0.0002 & 0 & -0.0006 \\ -0.0004 & 0 & 0.0011 \end{bmatrix}, \quad B_f=\begin{bmatrix} -0.0006 & 0.0006 \\ -0.0017 & 0.0015 \\ 0.1200 & 0 \\ -0.0018 & 0.1850 \\ -0.0034 & 0.0030 \\ 0.0016 & -0.0014 \\ -0.0029 & 0.0026 \end{bmatrix}$$

$$A_s=\begin{bmatrix} -0.4931 & 5.5763 & -5.3911 \\ -0.0650 & 0.6406 & -0.8540 \\ -0.1491 & 2.0413 & -2.0594 \end{bmatrix}$$

$$A_f=\begin{bmatrix} -1556.0000 & 0 & -0.0004 & 0.0002 & 0 & 0 & 33.5860 \\ 0 & -124.5000 & 212.6288 & 0.0007 & 112.6901 & 112.6892 & 0 \\ 0 & 0 & -3.3330 & 0 & 0 & 0 & 0 \\ 0 & 0 & -0.0013 & -32.4293 & 32.3041 & 32.3031 & 0 \\ 0 & 0 & -0.0026 & 331.8013 & -343.9998 & -341.0016 & 9.9042 \\ 0 & 221.9700 & 0.0013 & -0.0006 & -253.2001 & -254.8993 & 32.5260 \\ 8.1647 & -0.0057 & -0.3649 & 0.2151 & 5.6253 & 5.3949 & -13.6010 \end{bmatrix}$$

2.3.2 Eigenvalue Assignment for the Hydrogen Gas Reformer

It is known that the location of the closed-loop eigenvalues can shape the system transient response in a desired manner. Assume that we want to shape the transient response of the state space variables of the hydrogen gas reformer and place the closed-loop reformer slow eigenvalues at $-1, -2 \pm j2$ and the fast reformer eigenvalues at $-5, -7, -10 \pm j5, -15 \pm j10$. We have made this choice to keep three *open-loop* slow eigenvalues slow in the *closed-loop* feedback configuration and seven *open-loop* fast eigenvalues fast in the *closed-loop* feedback configuration, all for the purpose of testing the algorithm's numerical accuracy. Feedback should not mix state variables that are naturally slow and make them fast and the other way around. However, the newly proposed two-stage feedback design technique can handle *any desired choice of the closed-loop eigenvalues* under assumption that the system is controllable. To be able to assign the eigenvalues via linear feedback in any location, the hydrogen gas reformer must be controllable (Sinha, 2007; Chen, 2012). Importance of controllability for fuel cells has been considered in Radisavljevic (2011). It can be checked using MATLAB that the slow and fast models of the hydrogen gas reformer are both controllable, as well as the original hydrogen gas reformer model is also controllable.

According to the presented two-stage feedback design, the fast subsystem gain G_f is found first (Step 2) to locate $\lambda(A_f - B_f G_f) = \lambda_{fast}^{desired}$. This leads (with a help from the MATLAB function "place") to

$$G_f = 10^5 \begin{bmatrix} 1.2263 & 0.005 & -0.0104 & -0.0080 & 0.0197 & 0.0192 & -0.2864 \\ 0.4458 & -0.0312 & 0.0074 & -0.0419 & 0.0818 & 0.0815 & -0.1089 \end{bmatrix}$$

In Step 3, the Sylvester algebraic equation (2.30) is solved directly using the MATLAB function "lyap," which produced the accuracy of $O(10^{-7})$

$$P = 10^5 \begin{bmatrix} 2.3490 & 0.1284 & -0.0911 & 0.1198 & -0.2390 & -0.2394 & -0.5313 \\ 1.1541 & -0.0190 & 0.0021 & -0.0264 & 0.0586 & 0.0581 & -0.2720 \\ 2.5557 & -0.0452 & 0.0065 & -0.0618 & 0.1367 & 0.1355 & -0.6052 \end{bmatrix}$$

In Step 4, the slow subsystem feedback gain is found via the corresponding MATLAB function "place," such that the slow eigenvalues satisfy $\lambda\left(A_s - \left(B_{11} - PB_f\right)G_s\right) = \lambda_{slow}^{desired}$, which produced

$$G_s = \begin{bmatrix} 0.0042 & -0.0097 & 0.007 \\ 0.0015 & -0.0305 & 0.0144 \end{bmatrix}$$

It is important to emphasize that these slow and fast gains G_s, G_f are obtained using the reduced-order matrices only. The actual full-state feedback gain applied to the original hydrogen gas reformer using feedback from the original state space variables can be obtained in Step 5 as

$$G = \left[G_{1eq} \quad G_{2eq}\right] = \left[G_s + \left(G_f - \varepsilon G_s P\right)L \quad G_f - \varepsilon G_s P\right]$$

$$= 10^5 \begin{bmatrix} 1.2260 & 0.0001 & -0.0102 & -0.0084 & 0.0204 & 0.0200 & -0.2864 \\ 0.4430 & -0.0313 & 0.0075 & -0.0420 & 0.0819 & 0.0816 & -0.1083 \end{bmatrix}$$

It can be easily checked (using MATLAB) that indeed this gain produces the desired set of the eigenvalues with very high accuracy of $O(10^{-7})$. Namely, we have the following result

$$\lambda(A - BG) = \lambda_{system}^{desired} = \lambda_{slow}^{desired} \cup \lambda_{fast}^{desired} = \{-1.000000, -2.0000000 \pm j2.0000000\}$$

$$\cup\{-5.0000000, -7.0000000, -10.0000000 \pm 5.0000000, -12, -15.0000000 \pm j10.0000000\}$$

It should be emphasized that in this example, the eigenvalue assignment is performed in two stages by calculating independently slow and fast subsystem feedback gains using the reduced-order dynamic models and corresponding reduced-order matrices.

2.3.3 Optimal Slow and Eigenvalue Assigned Fast Subsystems

The methodology presented allows that different types of controllers can be designed for different subsystems. We can choose the same gain G_f to assign the hydrogen gas reformer closed-loop fast eigenvalues as in Sect. 2.3.2 at $-5, -7, -10 \pm j5, -15 \pm j10$ and the new gain G_s^{opt} to optimize its slow subsystem. The slow subsystem of the hydrogen gas reformer can be obtained from Step 4 of the two-stage design algorithm as

$$\frac{dx_s(t)}{dt} = A_s x_s(t) + B_s v(t), \quad B_s = B_{11} - PB_f \tag{2.40}$$

The slow subsystem optimized feedback gain is obtained by minimizing a quadratic performance criterion

$$J = \frac{1}{2}\int_0^\infty \left(x_s^T(t) R_1 x_s(t) + v^T(t) R_2 v(t)\right) dt \tag{2.41}$$

along the trajectories of the slow hydrogen gas reformer subsystem. Choosing the performance criterion penalty matrices as $R_1 = I_3$ and $R_2 = I_2$, we obtain the following expression for the optimal slow subsystem gain:

$$G_s^{opt} = \begin{bmatrix} 0.7413 & 0.2571 & 0.6139 \\ -0.6704 & 0.2607 & 0.6945 \end{bmatrix}$$

This gain will produce the optimal value for the performance criterion for the first (slow) subsystem as $J_{opt} = 0.8536$. Using the obtained values for G_s^{opt} in the expression for G_{eq}, as given in formula (2.34), produces the hybrid feedback gain as

$$G_{eq}^{hybrid} = \begin{bmatrix} G_{1eq}^{hybrid} & G_{2eq}^{hybrid} \end{bmatrix} = \begin{bmatrix} G_s^{opt} + \left(G_f - \varepsilon G_s^{opt} P\right) L & G_f - \varepsilon G_s^{opt} P \end{bmatrix}$$

This feedback gain optimizes the hydrogen gas reformer slow subsystem and simultaneously assigns its closed-loop fast subsystem eigenvalues in the desired locations at

$$-5, -7, -10 \pm j5, -15 \pm j10$$

2.4 Two-Stage Feedback Control of a Two-Time-Scale PEM Fuel Cell

Importance of fuel cells as green electric energy generators has been nicely demonstrated in the book by Hoffman and Dorgan (2012). PEM fuel cells are the most developed among all fuel cells, and they can be used for both mobile (vehicles, portable computing devices) and stationary applications (residential and industrial electric power generation and data centers). Moreover, the efficient use of renewable energy sources such as fuel cells for Internet Protocol (IP) over wavelength division multiplexing (WDM) fiber-optic networks with data centers was considered in Dong et al. (2012).

This section considers the eighth-order proton-exchange membrane (PEM) fuel cell mathematical model of Pukrushpan et al. (2004a, b) and shows that it has a two-time-scale property, indicating that the dynamics of three model state space variables operate in the slow time scale and the dynamics of five state variables operate in the fast time scale. This two-scale nature allows independent controllers to be designed in slow and fast time scales using only corresponding reduced-order slow (of dimension three) and fast (of dimension five) sub-models. The presented design facilitates the design of hybrid controllers, for example, the linear-quadratic optimal controller for the slow subsystem and the eigenvalue assignment controller for the fast subsystem. The design efficiency and its high accuracy are demonstrated via simulation on the considered PEM fuel cell mathematical model.

Fuel cells, due to their complex dynamics dictated by processes of different nature (electrochemical, thermodynamic, mechanical, chemical, electrical, etc.), offer many opportunities for control engineers. The proton-exchange membrane (PEM) is the most developed and the best understood class of fuel cells. Various PEM fuel cell dynamic models have been derived so far, and various controllers have been designed that work well for particular PEM models and under particular constraints and conditions; see, for example, Pukrushpan et al. (2004a, b), Tsourapas et al. (2007), Na et al. (2007), Na and Gou (2008), Nehrir and Wang (2009), Laurim et al. (2010), Gou et al. (2010), Matraji et al. (2013, 2015), Tong et al. (2013), Wang and Kim (2014), Wang and Guo (2015), and Laghrouche et al. (2015).

In this section we first show that the considered model PEM fuel cell, developed in Pukrushpan et al. (2004a, b), has the two-time-scale property (three state variable operate in the slow time scale and five state variables operate in the fast time scale). Then, we show how to design independent controllers in slow and fast time scales that take control of the slow and fast PEM fuel cell state variables. The obtained controllers are of the reduced-order and hence much simpler to design than the full-order controllers. Moreover, the presented methodology puts more light into the complex fuel cell dynamics and provides a better understanding of control design requirements for the considered class of PEM models.

2.4.1 PEM Fuel Cell Dynamics Mathematical Model

PEM fuel cells are devices that rely on the chemical production of electric energy and water. Within the fuel cell, a chemical reaction between hydrogen and oxygen takes place to form water and electricity. The fuel cell exploits the energy from this reaction, and uses it as a source of power for an electrical load, as schematically presented in Fig. 2.2.

Hydrogen gas enters into one side of the cell (known as the anode) and breaks apart into protons and electrons. This is accomplished with a reactive catalyst (usually platinum). The protons then move through a proton-exchange membrane that blocks electron flow and allows only hydrogen protons to pass through. This forces the electrons to travel through an external circuit, producing electricity for an external load. The electrons enter back into the cell at the opposite side of the membrane (known as the cathode).

It is here that oxygen gas (usually air) flows into the cell. Within the cathode, the free protons and electrons meet with the oxygen and react to produce water vapor. With the reaction completed, excess gases and water vapor flow out of the fuel cell. This process is described for a single fuel cell, but it can be extended to a stack of cells; see Fig. 2.3.

The process is the same for each individual cell but with the cells connected in series. Because of this connection, electron flow passes through each cell in series and then travels through an external path connected to opposite ends of the fuel cell

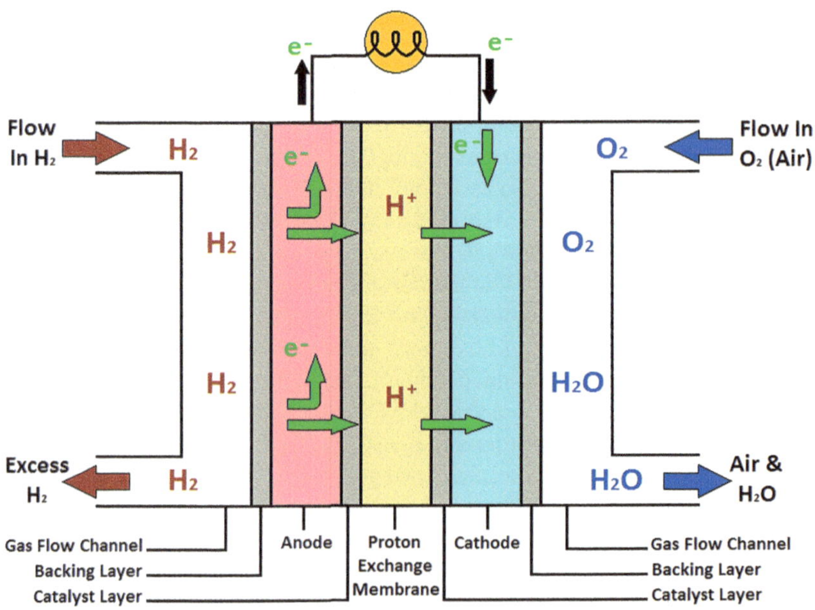

Fig. 2.2 PEM single cell reaction

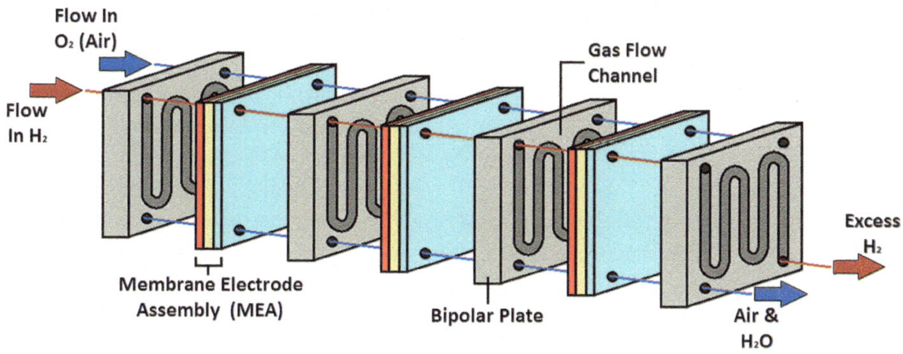

Fig. 2.3 PEM fuel cell stack

stack. The anode and cathode gas channels are kept separated and allow for inflow and outflow of gases through the fuel cell stack. The membrane electrode assembly (MEA) is the grouping of the anode, proton-exchange membrane, and cathode. Between the MEAs, gas channel plates are added to supply gas to each MEA. One side of the plate has channel flow for hydrogen gas, whereas the other side has channel flow for oxygen gas (air).

The linearized model of the polymer electrolyte membrane fuel cell has been developed by the University of Michigan researchers and considered in detail in a series of journal papers; see, for example, Pukrushpan et al. (2004a, b) and Tsourapas et al. (2007). The corresponding ninth-order nonlinear model of the considered PEM fuel cell and its state space variables are given by

$$\begin{aligned}\frac{dx(t)}{dt} &= f(x(t), u(t), w(t)) \\ x(t) &= \begin{bmatrix} x_1(t) & x_2(t) & x_3(t) & x_4(t) & x_5(t) & x_6(t) & x_7(t) & x_8(t) & x_9(t) \end{bmatrix} \\ &= \begin{bmatrix} m_{O_2}(t) & m_{H_2}(t) & m_{N_2}(t) & \omega_{cp}(t) & p_{sm}(t) & m_{sm}(t) & m_{H_2O_A}(t) & m_{H_2O_C}(t) & p_{rm}(t) \end{bmatrix}^T\end{aligned} \tag{2.42}$$

where $m_{O_2}, m_{H_2}, m_{N_2}, m_{H_2O_A}, m_{H_2O_C}$ are, respectively, masses of oxygen, hydrogen, nitrogen, anode side water vapor, and cathode side water vapor; ω_{cp} is the compressor (that blows the air (oxygen) on the cathode side) angular velocity; p_{sm} is supply manifold pressure and m_{sp} is the mass of gas in the supply manifold; p_{rm} is the return manifold pressure; $u(t) = v_{cm}(t)$ is the compressor motor voltage; and $w(t)$ is the disturbance and represents the fuel cell stack current, that is, $w(t) = I_{st}(t)$.

The output (measured) variables are given by

$$y(t) = \begin{bmatrix} W_{cp} & p_{sm} & v_{st} \end{bmatrix}^T = h_y(x, u, w) \tag{2.43}$$

where $W_{cp} = W_{cp}(x_4, x_5)$ is the compressor air molar flow rate, $p_{sm}(t) = x_5(t)$, and v_{st} is the fuel cell stack voltage. In addition to the measured output $y(t)$, the controlled output is given by

$$z(t) = \begin{bmatrix} e_{Pnet} & \lambda_{O_2} \end{bmatrix}^T = h_z(x, u, w) \tag{2.44}$$

where two controlled variables are $e_{Pnet} = P_{net}^{ref} - P_{net}$, the difference between the desired $P_{net} = 40$ kW and actual net power, and $\lambda_{O_2} = W_{O_2}^{in} / W_{O_2}^{reacted}$ known as the oxygen excess ratio.

The linearized system matrices at steady state (corresponding to the following steady-state values: $P_{net}^{ss} = 40$ kW,$I^{ss} = 191$ A,$v_{cm}^{ss} = 164$ V,$\lambda_{O_2}^{ss} = 2$) can be found in Pukrushpan et al. (2004a), p. 145. The data from Pukrushpan et al. (2004a) are represented using corresponding state space matrices with the state space model of the linearized system given by

$$\begin{aligned}
&\frac{\delta x(t)}{dt} = A\delta x(t) + B\delta u(t) + F\delta w(t) \\
&\delta y(t) = C_y \delta x(t) + D_{yu}\delta u(t) + D_{yw}\delta w(t) \\
&\delta z(t) = C_z \delta x(t) + D_{zu}\delta u(t) + D_{zw}\delta w(t) \\
&x(t) = x_{ss} + \delta x(t), \quad u(t) = u_{ss} + \delta u(t), \quad y(t) = y_{ss} + \delta y(t) \\
&z(t) = z_{ss} + \delta z(t), \quad w(t) = w_{ss} + \delta w(t)
\end{aligned} \tag{2.45}$$

where $\delta z(t) = \begin{bmatrix} \delta e_{P_{net}}(t) & \delta\lambda_{O_2}(t) \end{bmatrix}^T$. It is assumed, for the reason of design simplicity, that $\delta w(t) = \delta I_{st}(t) = 0$ so that in this section we consider the model

$$\begin{aligned}
&\frac{\delta x(t)}{dt} = A\delta x(t) + B\delta u(t) \\
&\delta y(t) = C_y \delta x(t) + D_{yu}\delta u(t) \\
&\delta z(t) = C_z \delta x(t) + D_{zu}\delta u(t) \\
&z(t) = z_{ss} + \delta z(t), \quad x(t) = x_{ss} + \delta x(t) \\
&u(t) = u_{ss} + \delta u(t), \quad y(t) = y_{ss} + \delta y(t)
\end{aligned} \tag{2.46}$$

Note that the original model (2.42) is of order nine. It is interesting to observe that the obtained linearized model is of order eight, Pukrushpan et al. (2004a, b). Namely, the state variable that has a negligible impact on the system dynamics (mass of water in the cathode) is removed from the original ninth-order linearized model using the results of Zhou and Doyle (1998), obtained by studying the singular values of the corresponding observability and controllability Gramians that serve as observability and controllability measures (Zhou and Doyle 1998). As a matter of fact, the system models that are both weakly controllable and weakly observable can be neglected, which can be achieved using the balancing transformation that brings the system into the coordinates in which the controllability and observability Gramians are identical and diagonal (balanced realization) (Zhou and Doyle 1998). Controllability and observability issues of a PEM fuel cell were considered in several papers; see, for example, Serra et al. (2005) and Radisavljevic (2011).

The corresponding numerical value for the linearized model taken from Pukrushpan et al. (2004a, b) is given by

$$A = \begin{bmatrix} -6.30908 & 0 & -10.9544 & 0 & 83.74458 & 0 & 0 & 24.05866 \\ 0 & -161.083 & 0 & 0 & 51.52923 & 0 & -18.0261 & 0 \\ -18.7858 & 0 & -46.3136 & 0 & 275.6592 & 0 & 0 & 158.3741 \\ 0 & 0 & 0 & -17.3506 & 193.9373 & 0 & 0 & 0 \\ 1.299576 & 0 & 2.969317 & 0.3977 & -38.7024 & 0.105748 & 0 & 0 \\ 16.64244 & 0 & 38.02522 & 5.066579 & -479.384 & 0 & 0 & 0 \\ 0 & -450.386 & 0 & 0 & 142.2084 & 0 & -80.9472 & 0 \\ 2.02257 & 0 & 4.621237 & 0 & 0 & 0 & 0 & -51.2108 \end{bmatrix}$$

$$B = \begin{bmatrix} 0 & 0 & 0 & 3.946683 & 0 & 0 & 0 & 0 \end{bmatrix}^T$$

$$C_y = \begin{bmatrix} 0 & 0 & 0 & 5.066579 & -116.446 & 0 & 0 & 0 \\ 0 & 0 & 0 & 0 & 1 & 0 & 0 & 0 \\ 12.96989 & 10.32532 & -0.56926 & 0 & 0 & 0 & 0 & 0 \end{bmatrix}$$

$$C_z = \begin{bmatrix} -2.4837 & -1.9773 & 0.109013 & -0.21897 & 0 & 0 & 0 & 0 \\ -0.63477 & 0 & -1.45035 & 0 & 13.84308 & 0 & 0 & 0 \end{bmatrix}$$

$$D_{zu} = \begin{bmatrix} 0.169141 \\ 0 \end{bmatrix}, \quad D_{yu} = \begin{bmatrix} 0 \\ 0 \\ 0 \end{bmatrix}$$

2.4.2 *Two-Time-Scale Structure of the PEM Fuel Cell*

Finding the eigenvalues of the PEM fuel cell system matrix A, it can be seen that the eigenvalues are widely spread so that they can be clustered into two groups, three small eigenvalues close to the imaginary axis (slow eigenvalues) and five large (in absolute value) eigenvalues far from the imaginary axis (fast eigenvalues), that is,

$$\lambda_1 = -1.4038, \quad \lambda_2 = -1.6473, \quad \lambda_3 = -2.9151$$
$$\lambda_4 = -18.2582, \quad \lambda_5 = -22.4040, \quad \lambda_6 = -46.1768, \quad \lambda_7 = -89.4853, \quad \lambda_8 = -219.6262$$

This clearly indicates the two-time-scale structure of the considered PEM fuel cell model. As a matter of fact, the PEM fuel cell eigenvalues can be clustered in three and even four groups indicating the fuel cell operation in three or even four time scales. For the purpose of simplicity, and for the purpose of using the available tools (the two-stage feedback controller design technique), we will limit our attention in this study only to two time scales with three slow and five fast eigenvalues.

The three- and four-stage designs of linear feedback controllers will be considered in the follow-up chapters. Moreover, discrete-time versions of these designs will be also needed for digital controller implementations and PC (personal computer) in the loop control of PEM fuel cells. The discrete-time *two-stage* two-time-scale design of feedback controllers for linear dynamic systems will be presented in detail in Chap. 3. The main results and ideas for the corresponding discrete three-time *three-stage* feedback design will be considered in Chap. 5.

Linear systems with slow and fast modes (singularly perturbed systems; Kokotovic et al. (1999), Naidu and Calise (2001), Kuehn (2015)) are particularly well suited for the considered two-stage feedback design. For this class of systems, in general, numerical ill-conditioning appears if one attempts to design a linear feedback controller using the entire (full-order) system. Singularly perturbed systems have numerous applications in all areas of engineering and sciences (Naidu and Calise 2001). In this section, further design simplifications will be achieved by specializing the proposed design from Sect. 2.2 to singularly perturbed linear systems so that only solutions of linear *algebraic equations* will be required.

In practice, usually the small singular perturbation parameter ε is determined as a ratio of the magnitudes of the fastest slow eigenvalue and the slowest fast eigenvalue, that is, in the case of the considered PEM fuel cell, it is given by

$$\varepsilon = \frac{|\lambda_{s\max}|}{|\lambda_{f\min}|} = \frac{2.9151}{18.2582} = 0.1597 \tag{2.47}$$

The PEM fuel cell mathematical model can be put into the explicit singularly perturbed form by exchanging the rows and columns in the original system matrix. The PEM fuel cell mathematical model as defined in formula (2.46), with the corresponding matrices known, is in the implicit singularly perturbed form, namely, the system matrix A has the form $A = A(\varepsilon)$. To get the *explicit* singularly perturbed form consistent with (2.20), we have to exchange the order of the state space variables such that the matrix A_{22} corresponding to the fast variables is nonsingular. In general, this is not an easy task, especially for higher-dimensional systems. After several attempts, we have found that the exchange of state variables using a similarity matrix V obtained as a product of permutation matrices given by $V = I_{25}I_{24}I_{27}I_{15}I_{36}$ will produce the desired singularly perturbed form defined in (2.20), as originally presented in Radisavljevic-Gajic et al. (2015). A permutation matrix I_{ij} is obtained from the identity matrix by interchanging rows i and j (Golub and Van Loan 2012). The obtained matrix V is given by

$$V = \begin{bmatrix} 0 & 0 & 0 & 0 & 1 & 0 & 0 & 0 \\ 1 & 0 & 0 & 0 & 0 & 0 & 0 & 0 \\ 0 & 0 & 0 & 0 & 0 & 1 & 0 & 0 \\ 0 & 0 & 0 & 0 & 0 & 0 & 1 & 0 \\ 0 & 0 & 0 & 1 & 0 & 0 & 0 & 0 \\ 0 & 0 & 1 & 0 & 0 & 0 & 0 & 0 \\ 0 & 1 & 0 & 0 & 0 & 0 & 0 & 0 \\ 0 & 0 & 0 & 0 & 0 & 0 & 0 & 1 \end{bmatrix} \tag{2.48}$$

The similarity transformation $A_{sp} = VAV^T$(note that $V^T = V^{-1}$) produces the explicit singularly structure for the considered PEM fuel cell mathematical model given by

$$\begin{aligned} A_{sp} &= VAV^T = \begin{bmatrix} A_{11sp} & A_{12sp} \\ A_{21sp} & A_{22sp} \end{bmatrix} \\ &= \begin{bmatrix} -38.7024 & 1.2996 & 0.1057 & 0 & 0.3977 & 2.9693 & 0 & 0 \\ 83.7446 & -6.3091 & 0 & 0 & 0 & -10.9544 & 0 & 24.0587 \\ -479.3840 & 16.6424 & 0 & 0 & 5.0666 & 38.0252 & 0 & 0 \\ 142.2084 & 0 & 0 & -80.9472 & 0 & 0 & -450.3860 & 0 \\ 193.9373 & 0 & 0 & 0 & -17.3506 & 0 & 0 & 0 \\ 275.6592 & -18.7858 & 0 & 0 & 0 & -46.3136 & 0 & 158.3741 \\ 51.5292 & 0 & 0 & -18.0261 & 0 & 0 & -161.3136 & 0 \\ 0 & 2.2026 & 0 & 0 & 0 & 46.6212 & 0 & -51.2108 \end{bmatrix} \end{aligned} \tag{2.49}$$

The matrices in (2.20) and (2.49) are related by

$$A_{11} = A_{11sp}, \quad A_{12} = A_{12sp}, \quad A_{21} = \varepsilon A_{21sp}, \quad A_{22} = \varepsilon A_{22sp} \tag{2.50}$$

What has been done in (2.49) is equivalent to ordering of the original state space variables as

$$[x_5(t) \quad x_1(t) \quad x_6(t) \quad x_7(t) \quad x_4(t) \quad x_3(t) \quad x_2(t) \quad x_8(t)] \tag{2.51}$$

With this ordering of the state space variables, Assumption 2.1 is satisfied, the corresponding slow and fast subsystems (to be defined) are controllable, and the slow subsystem contains all slow eigenvalues, and the fast subsystem contains all fast eigenvalues. With the slow-fast ordering of the state variables as in (2.51), we conclude that $x_5(t), x_1(t), x_6(t)$ are the slow state variables and that the remaining state space variables $x_7(t), x_4(t), x_3(t), x_2(t), x_8(t)$ are the fast state variables.

Comment 2.1 It is interesting to observe from (2.51) that the pressure of oxygen, $x_1(t)$, is in the subsystem representing the slow variables and the pressure of hydrogen, $x_2(t)$, is in the subsystem representing the fast variables. Another interesting observation, even though expected from the physical point of view, can be mathematically verified by using the balancing transformation (Zhou and Doyle 1998) and finding the measures of the dynamic dominance of each state space variable in the considered singularly perturbed system. Namely, the balancing transformation reveals that the most dominant variables in this system (the variables with the highest energy) are $x_1(t)$ and $x_2(t)$. What is interesting here is pointing out the fact that $x_2(t)$ is dominant despite being fast (usually fast state variables are low-energy signals, hence nondominant signals).

The input and output matrices in the explicit singularly perturbed form of the PEM fuel cell are

$$B_{sp} = \begin{bmatrix} B_{11sp} \\ B_{22sp} \end{bmatrix} = VB = \begin{bmatrix} 0 & 0 & 0 & 0 & 3.9467 & 0 & 0 & 0 \end{bmatrix}^T \tag{2.52}$$

$$\begin{aligned} C_{sp} &= \begin{bmatrix} C_{11sp} & C_{22sp} \end{bmatrix} = C_y V^T \\ &= \begin{bmatrix} -116.4460 & 0 & 0 & 0 & 5.0666 & 0 & 0 & 0 \\ 1 & 0 & 0 & 0 & 0 & 0 & 0 & 0 \\ 0 & 12.9689 & 0 & 0 & 0 & -0.5693 & 10.3253 & 0 \end{bmatrix} \end{aligned} \tag{2.53}$$

The system input matrices of (2.20) and (2.52) are related by the following relationships

$$B_{11} = B_{11sp}, \quad B_{22} = \varepsilon B_{22sp} \tag{2.54}$$

Now we apply the two-stage two-time-scale feedback controller design from Sect. 2.2 to the proton-exchange membrane fuel cell model defined in this section. To perform linear feedback design in two stages for singularly perturbed linear systems (including the eigenvalue assignment), it is needed to find solutions of the nonlinear algebraic equation (2.21) and the linear algebraic equation (2.30). A unique solution of (2.21) exists for sufficiently small values of ε under Assumption 2.1. In that case, the algebraic equation (2.21) can be solved by performing fixed-point iterations on a system of *linear algebraic equations* as shown in (2.22). Moreover, if ε is not sufficiently small, the eigenvector method (Medanic 1982) can be used to solve (2.21). Under special assumptions, the Newton method can be also used for solving (2.21). Having obtained L from (2.21), the algebraic equation (2.30) can be also solved iteratively for sufficiently small values of ε as a system of *linear algebraic equations* using the following iteration

$$\begin{aligned} P^{(i+1)}\left(A_f - B_f G_f\right) &= A_{12} - B_{11} G_f + \varepsilon A_s P^{(i)} = 0 \\ P^{(0)} &= \left(A_{12} - B_{11} G_f\right)\left(A_f - B_f G_f\right)^{-1} \end{aligned} \tag{2.55}$$

Note that $A_f - B_f G_f$ is the closed-loop fast subsystem matrix and hence asymptotically stable (Chen 2012) so that $A_f - B_f G_f$ is invertible. If ε is not sufficiently small, equation (2.30) should be solved directly as the algebraic Sylvester equation, in which case the unique solution is obtained under the assumption that matrices εA_s and $A_f - B_f G_f$ have no eigenvalues in common (Chen 2012), which is always true for singularly perturbed systems since they have eigenvalues separated into two clusters: the slow ones close to the imaginary axis and the fast ones far from the imaginary axis. Moreover, the slow eigenvalues are multiplied by a small positive parameter ε (due to the well-known result from linear algebra that holds for the eigenvalues, $\lambda(\varepsilon A_s) = \varepsilon\lambda(A_s)$; Stewart (1973)) so that the unique solution of the algebraic Sylvester equation (2.30) always exists.

Comment 2.2 Despite of the fact that the Sylvester equation (2.30) is a linear algebraic equation, in some cases, the direct methods for its numerical solution cannot produce very high accuracy as was pointed in Gao and Bai (2010); see also Simoncini (2016). It is possible that for some systems the MATLAB function "lyap" (used also for solving the Sylvester algebraic equation) produces only a limited accuracy for the Sylvester equation. In such cases, solving the Sylvester algebraic equation (2.30) iteratively as presented in (2.55) can produce very high accuracy (unless the small parameter ε is not sufficiently small, in which case the iterations (2.55) do not converge).

The expression for the slow and fast system output matrices (even they are not directly involved in the two-stage feedback design procedures) can be obtained using the relations between the original and slow-fast coordinates. These relationships are given by

$$\begin{aligned}\begin{bmatrix} x_s(t) \\ x_f(t) \end{bmatrix} &= T_{2sp}T_{1sp}\begin{bmatrix} x_{\mathrm{I}}(t) \\ x_{\mathrm{II}}(t) \end{bmatrix} = \begin{bmatrix} I & -\varepsilon P \\ 0 & I \end{bmatrix}\begin{bmatrix} I & 0 \\ L & I \end{bmatrix}\begin{bmatrix} x_{\mathrm{I}}(t) \\ x_{\mathrm{II}}(t) \end{bmatrix} \\ &= \begin{bmatrix} I - \varepsilon PL & -\varepsilon P \\ L & I \end{bmatrix}\begin{bmatrix} x_{\mathrm{I}}(t) \\ x_{\mathrm{II}}(t) \end{bmatrix}\end{aligned} \tag{2.56}$$

and

$$\begin{aligned}\begin{bmatrix} x_{\mathrm{I}}(t) \\ x_{\mathrm{II}}(t) \end{bmatrix} &= T_{1sp}^{-1}T_{2sp}^{-1}\begin{bmatrix} x_s(t) \\ x_f(t) \end{bmatrix} = \begin{bmatrix} I & 0 \\ -L & I \end{bmatrix}\begin{bmatrix} I & \varepsilon P \\ 0 & I \end{bmatrix}\begin{bmatrix} x_s(t) \\ x_f(t) \end{bmatrix} \\ &= \begin{bmatrix} I & \varepsilon P \\ -L & I - \varepsilon LP \end{bmatrix}\begin{bmatrix} x_s(t) \\ x_f(t) \end{bmatrix}\end{aligned} \tag{2.57}$$

Due to the coordinate changes of the state space equations, the output of the singularly perturbed system defined in (2.20) is also changed, and, in the new coordinates, it is given by

$$
\begin{aligned}
y(t) &= \begin{bmatrix} C_{11sp} & C_{22sp} \end{bmatrix} \begin{bmatrix} x_{\mathrm{I}}(t) \\ x_{\mathrm{II}}(t) \end{bmatrix} = \begin{bmatrix} C_{11sp} & C_{22sp} \end{bmatrix} \begin{bmatrix} I & \varepsilon P \\ -L & I - \varepsilon LP \end{bmatrix} \begin{bmatrix} x_s(t) \\ x_f(t) \end{bmatrix} \\
&= \begin{bmatrix} C_{11sp} - C_{22sp}L & C_{22sp} + \varepsilon\left(C_{11sp} - C_{22sp}L\right)P \end{bmatrix} \begin{bmatrix} x_s(t) \\ x_f(t) \end{bmatrix} = C_s x_s(t) + C_f x_f(t)
\end{aligned}
\tag{2.58}
$$

with

$$
C_s = C_{11sp} - C_{22sp}L, \quad C_f = C_{22sp} + \varepsilon C_s P \tag{2.59}
$$

2.4.3 PEM Fuel Cell Slow-Fast Two-Stage Controller Design Simulation

In this section, the controller design is performed in two stages by calculating independently slow and fast subsystem feedback gains using the reduced-order dynamic models and corresponding reduced-order matrices of the considered PEM fuel cell mathematical model. We first perform the design using the eigenvalue assignment methodology for both the slow and fast subsystems, and then, we design a mixed-type controller, a hybrid controller, by assigning the desired eigenvalues only to the fast subsystem and using the linear-quadratic optimal controller for the slow subsystem. Of course, we could perform the linear-quadratic optimal controller design for both slow and fast subsystems or use any other linear feedback controller for either of the subsystems.

Using the value for the small singular perturbation parameter given in (2.47) and the system model matrix presented (2.49) (see also (2.50)), we have the following submatrices corresponding to the singularly perturbed structure defined in (2.20)

$$
A_{11} = \begin{bmatrix} -38.7024 & 1.2996 & 0.1057 \\ 83.7446 & -6.3091 & 0 \\ -479.3840 & 16.6424 & 0 \end{bmatrix}, \quad A_{21} = \begin{bmatrix} 22.3267 & 0 & 0 \\ 30.4482 & 0 & 0 \\ 43.2785 & -2.9494 & 0 \\ 8.0901 & 0 & 0 \\ 0 & 0.3175 & 0 \end{bmatrix}
$$

$$
A_{12} = \begin{bmatrix} 0 & 0.3977 & 2.9693 & 0 & 0 \\ 0 & 0 & -10.9544 & 0 & 24.0587 \\ 0 & 5.0666 & 38.0252 & 0 & 0 \end{bmatrix}
$$

$$A_{22} = \begin{bmatrix} -12.7087 & 0 & 0 & -70.7106 & 0 \\ 0 & -2.7240 & 0 & 0 & 0 \\ 0 & 0 & -7.2712 & 0 & 24.8647 \\ -2.8301 & 0 & 0 & -25.2900 & 0 \\ 0 & 0 & 0.7255 & 0 & -8.0401 \end{bmatrix}$$

The solution of the L-equation is found by running the Newton method iterations for $k = 6$, which produced the accuracy of $E^{(k)} = O(10^{-13})$, with the accuracy being defined by a matrix norm of the following expression

$$E^{(k)} = \varepsilon L^{(k)} A_{11} - A_{22} L^{(k)} - \varepsilon L^{(k)} A_{12} L^{(k)} + A_{21}$$

Having obtained the values for $L \approx L^{(6)}$, matrices A_s, A_f, B_f are found from formula (2.25) as

$$A_s = \begin{bmatrix} -3.5199 & 0.0291 & 0.0429 \\ -2.3591 & -1.6426 & 0.1232 \\ -29.2031 & 0.3737 & -0.8037 \end{bmatrix}$$

$$L = \begin{bmatrix} 0.1933 & -0.0012 & -0.0017 \\ -13.9420 & 0.0250 & 0.0360 \\ -9.9813 & 0.4245 & 0.0163 \\ -0.3491 & 0.0002 & 0.0003 \\ -0.9658 & -0.0007 & 0.0023 \end{bmatrix}$$

and

$$A_f = \begin{bmatrix} -12.7087 & 0.0170 & 0.0820 & -70.7106 & -0.0044 \\ 0 & -3.5660 & -6.3278 & 0 & 0.0944 \\ 0 & -0.6102 & -12.5569 & 0 & 26.4682 \\ -2.8301 & -0.0216 & -0.1614 & -25.2900 & 0.0007 \\ 0 & -0.0585 & 0.2903 & 0 & -8.0427 \end{bmatrix}, \quad B_f = \begin{bmatrix} 0 \\ 0.6196 \\ 0 \\ 0 \\ 0 \end{bmatrix}$$

Note that the matrix $B_s = B_s(G_f)$ depends on the fast subsystem feedback gain; see (2.40) used for the first subsystem so it will be determined in the subsequent part of this section when we define and find the corresponding feedback controller.

Eigenvalue Assignment for the PEM Fuel Cell

Assume that for some reason (e.g., to achieve the required transient response), we intend to place the PEM fuel cell closed-loop slow eigenvalues at $-1, -1.5 \pm j1$ and the closed-loop fast eigenvalues at $-9 \pm j5, -10, -11, -12$. The considered two-stage feedback design technique can handle *any desired choice of the closed-loop eigenvalues* under the assumption that the system is controllable. It can be checked using MATLAB that both the slow and fast models of the PEM fuel cell are both controllable, as well as the original PEM fuel cell model is also controllable.

The fast subsystem gain G_f is found first (Step 2) via the eigenvalue assignment technique to locate $\lambda\left(A_f - B_f G_f\right) = \lambda_{fast}^{desired}$. Using the MATLAB function "place," we got

$$G_f = 10^5\left[-1.898 \quad -0.0090 \quad -0.0018 \quad -1.4579 \quad -0.0298\right]$$

The Sylvester algebraic equation is solved directly using the MATLAB function "lyap," with the accuracy of $O(10^{-7})$ as

$$P = 10^7 \times \begin{bmatrix} -0.0176 & -0.0001 & -0.0010 & -0.1333 & -0.0087 \\ -0.0344 & -0.0001 & -0.0020 & -0.2608 & -0.0169 \\ -0.2321 & -0.0007 & -0.0135 & -1.7570 & -0.1151 \end{bmatrix}$$

The matrix B_s obtained from (2.40) is given by

$$B_s = 10^3 \begin{bmatrix} 0.3124 \\ 0.6136 \\ 4.1177 \end{bmatrix}$$

In Step 4, the slow subsystem feedback gain is found such that $\lambda(A_s - B_s G_s) = \lambda_{slow}^{desired}$ producing

$$G_s = \left[-0.0615 \quad 0.0038 \quad 0.0036\right]$$

Both slow and fast gains G_s, G_f are calculated using only the reduced-order matrices. The actual full-state feedback gain applied to the original PEM fuel cell model is calculated as

$$\begin{aligned} G_{eq} &= \left[G_{1eq} \quad G_{2eq}\right] = \left[G_s + \left(G_f - \varepsilon G_s P\right)L \quad G_f - \varepsilon G_s P\right] \\ &= 10^5\left[0.5350 \quad -0.0008 \quad -0.0002 \quad -0.1915 \quad -0.0009 \quad -0.0017 \quad -1.4711 \quad -0.0307\right] \end{aligned}$$

It can be easily checked via MATLAB that this gain provides the desired slow and fast eigenvalues with high accuracy of $O(10^{-6})$, that is

$$\begin{aligned} &\lambda(A - BG) = \lambda_{system}^{desired} = \lambda_{slow}^{desired} \cup \lambda_{fast}^{desired} = \{-1.000000 \quad -1.500000 \pm j1.000000\} \\ &\cup\{-9.000000 \pm j5.000000 \quad -10.000000 \quad -11.000000 \quad -12.000000\} \end{aligned}$$

Note that the MATLAB function "lyap" produced accuracy for the matrix P of only $O(10^{-7})$.

Comment 2.3

It should be noticed that some elements, the feedback gain G_f and matrix P, contain large values. This is the consequence of the fact that the fast state variables are very weakly controllable (have very small Hankel singular values (Zhou and Doyle 1998), which is equivalent to the corresponding controllability Gramian being very close to singular). It is well known that such systems are very difficult to control and large efforts, high gain signals, are needed.

Hybrid Optimal Slow-Fast Eigenvalue Assigned PEM Fuel Cell Controller
The two-stage design facilitates that different types of controllers can be designed for each subsystem. In this subsection, we design the gain G_f to assign the PEM fuel cell closed-loop fast eigenvalues like in Sect. 2.4.3 at the desired locations, say $-9 \pm j5$, $-10, -11, -12$, and the gain G_s^{opt} to optimize the slow subsystem. The slow subsystem of the considered PEM fuel cell is defined in Step 4 by matrices A_s and $B_s = B_{11} - PB_f$ as

$$\frac{dx_s(t)}{dt} = A_s x_s(t) + B_s v(t)$$

The slow subsystem is optimized in the sense that a quadratic performance criterion

$$J_s = \frac{1}{2} \int_0^\infty \left(x_s^T(t) R_1 x_s(t) + v^T(t) R_2 v(t) \right) dt$$

is minimized along the trajectories of the slow PEM fuel cell subsystem. The performance criterion penalty matrices are chosen as $R_1 = I_3$ and $R_2 = I_2$, which produce

$$G_s^{opt} = \begin{bmatrix} 0.0965 & 0.1875 & 0.9779 \end{bmatrix}$$

This gain provides the optimal value for the performance criterion for the first (slow) subsystem as $J_{opt} = 0.4783$. The selected gains G_f and G_s^{opt} produce the value for the equivalent full-state feedback for the original PEM fuel cell given by

$$\begin{aligned} G_{eq}^{opt} &= \begin{bmatrix} G_{1eq} & G_{2eq} \end{bmatrix} = \begin{bmatrix} G_s^{opt} + \left(G_f - \varepsilon G_s^{opt} P \right) L & G_f - \varepsilon G_s^{opt} P \end{bmatrix} \\ &= 10^6 \times \begin{bmatrix} -0.8291 & -0.0091 & 0.0003 & 0.3502 & 0.0010 & -0.0216 & 2.6487 & 0.1800 \end{bmatrix} \end{aligned}$$

This *hybrid gain optimizes the PEM fuel cell slow subsystem and at the same time assigns the closed-loop fast subsystem eigenvalues in the desired locations* $-9 \pm j5, -10, -11, -12$.

2.5 PEM Fuel Cell Observer Design

With the two-stage state feedback control design completed, the fuel cell system controller implementation requires knowledge of all eight state space variables since $u(t) = u(x(t)) = -G_{eq}x(t)$. However, the output of the system, represented by $y(t)$, directly measures only three of the system states. The solution to this problem is to use an observer to estimate all of the system state space variables. Observers are used in many areas of sciences and engineering (Sinha 2007; Chen 2012; Radisavljevic-Gajic 2015c; Ali et al. 2015). The use of observers in control of PEM fuel cells was considered in Pilloni et al. (2015).

An observer is designed as a dynamic system of the same order and the same structure as the original system. The goal that the difference between the actual system output and the observer output, known as the system output observation error, tends toward zero in time, which implies that the observer state variables tend to the actual system state space variables.

For the linearized fuel cell model considered in this chapter with $D_{yu} = 0$, we have

$$\begin{aligned} \frac{\delta x(t)}{dt} &= A\delta x(t) + B\delta u(t) \\ \delta y(t) &= C_y \delta x(t) \\ x(t) &= x_{ss} + \delta x(t), \quad u(t) = u_{ss} + \delta u(t), \quad y(t) = y_{ss} + \delta y(t) \end{aligned} \tag{2.60}$$

An observer is defined by the following differential equation (Sinha 2007; Chen 2012; Radisavljevic-Gajic 2015c)

$$\begin{aligned} \frac{\delta \hat{x}(t)}{dt} &= A\delta \hat{x}(t) + B\delta u(t) + K(\delta y(t) - \delta \hat{y}(t)) \\ &= (A - KC_y)\delta \hat{x}(t) + B\delta u(t) + K\delta y(t) \\ \delta \hat{y}(t) &= C_y \delta \hat{x}(t) \end{aligned} \tag{2.61}$$

where $\delta x(t)$ are the variations of the actual system states around its nominal values, $\delta \hat{x}(t)$ are the observer estimated state variations, and $\delta u(t)$ are variations of the system input, and K is the observer gain.

The observer error variations are defined as

$$\delta e(t) = \delta x(t) - \delta \hat{x}(t) \tag{2.62}$$

which leads to the differential equation for dynamics of the observation error variations

$$\frac{\delta e(t)}{dt} = (A - KC_y)\delta e(t) \tag{2.63}$$

The observer gain K has to be chosen such that the observer feedback matrix $A - KC_y$ is asymptotically stable. That can be achieved by choosing the observer eigenvalues and placing them in the left-half complex plane. The observer gain can be found with a help from the MATLAB function "place" that finds the observer gain K such that the eigenvalues are placed in the desired locations. Note that the locations of the observer eigenvalues should be chosen such that the closed-loop observer is considerably faster than the closed-loop system (the speed is determined by the locations of the system closed-loop eigenvalues; the eigenvalues of the matrix $A - BG_{eq}$, where G_{eq} is the linear proportional full-state feedback gain). In practice, this is accomplished by placing the observer eigenvalues $\lambda_{obs} = \lambda(A - KC_y)$ to be around five to six times faster than the fuel cell system eigenvalues $\lambda_{system}^{desired} = \lambda(A - BG_{eq})$. As long as the matrix pair (A, C_y) is observable, the observer eigenvalues can be placed at any desired location (Sinha 2007; Chen 2012; Radisavljevic-Gajic 2015c). In the case of this PEM fuel cell, the system is observable.

The observer is implemented using the SIMULINK state space block as a system with one vector input and one vector output (Radisavljevic-Gajic 2015c), that is, as

$$\frac{\delta\hat{x}(t)}{dt} = (A - KC_y)\delta\hat{x}(t) + [B \quad K]\begin{bmatrix}\delta u(t)\\ \delta y(t)\end{bmatrix} \tag{2.64}$$

Figure 2.4 shows a Simulink model block diagram for both the fuel cell system and observer. State feedback control used with the estimated states is $\delta u(t) = -G_{eq}\delta\hat{x}(t)$, with G_{eq} found in Sect. 2.4.3. That gain was found in that section such that the slow closed-loop fuel cell system eigenvalues are placed at $-1, -1.5 \pm j1$ and the closed-loop fast fuel cell eigenvalues are placed at $-9 \pm j5, -10, -11, -12$. The observer eigenvalues are chosen such that the observer eigenvalues are five to six times faster than the fuel cell eigenvalues. The observer eigenvalues are placed at

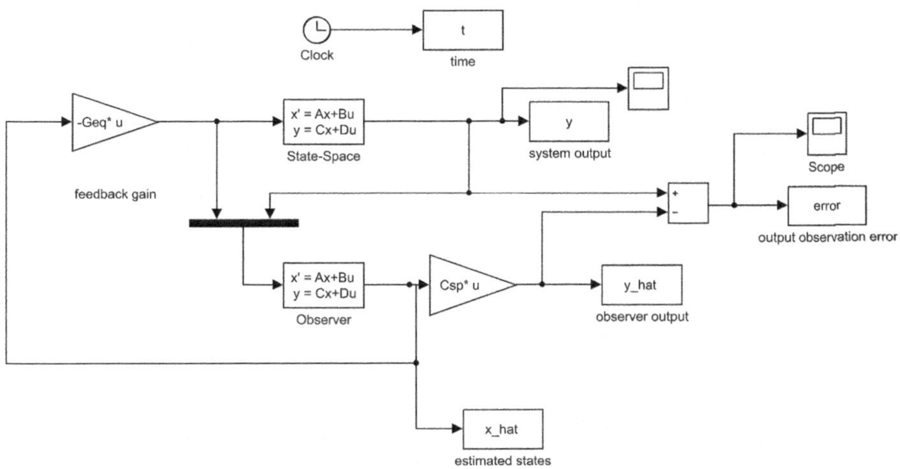

Fig. 2.4 Simulink model of the fuel cell system and the observer

$$\lambda_{obs} = \lambda(A - KC_y) = \{-60 \pm j10, -62, -64, -66, -68 \pm j20, -70\}$$

Using the "place" function in MATLAB, the observer gain matrix K is obtained as

$$K = 10^4 \begin{bmatrix} 0.0000 & 0.0074 & 0.0000 \\ 0.0000 & -0.0271 & 0.0003 \\ -0.1242 & -3.2463 & 0.0151 \\ 0.0007 & -0.2496 & -0.0025 \\ 0.0009 & 0.1266 & 0.0000 \\ 0.0024 & 0.2121 & -0.0004 \\ 0.0011 & -0.0811 & -0.0010 \\ 0.0004 & 0.0256 & 0.0000 \end{bmatrix}$$

Having obtained the observer gain matrix K, the Simulink model can be run using the observer estimated states as the full-state feedback input into the system. The observer gain is relatively high since the observability Gramian matrix is close to singular (it has small Hankel singular values; Zhou and Doyle 1998), which indicates that relatively large efforts are needed to observe all state variables of the considered PEM fuel cell. Despite that fact, the observation error defined as the difference between the actual and observed fuel cell outputs converges to zero rather quickly within 0.2 s, which is pretty remarkable; see Fig. 2.5.

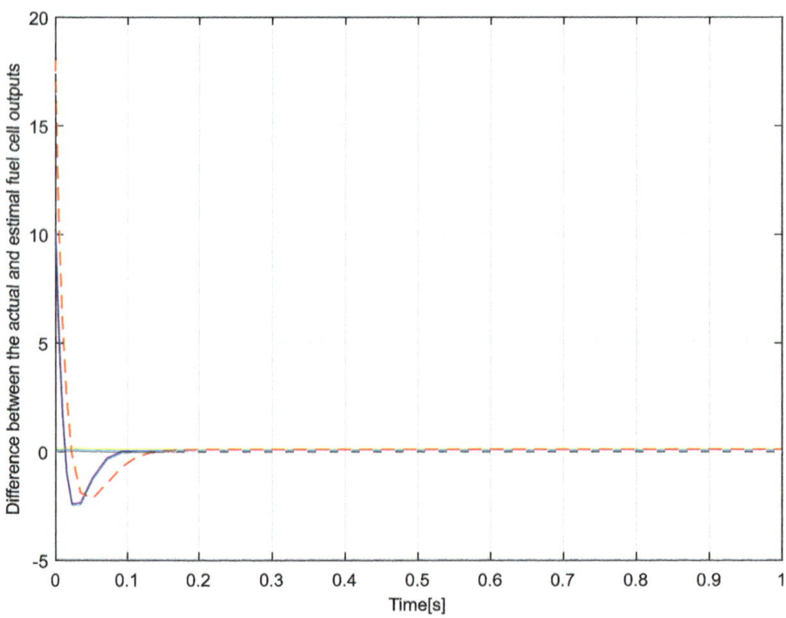

Fig. 2.5 Errors between the actual and observed (estimated) outputs

It should be emphasized that the observer design presented uses the classical method of Sinha (2007), Chen (2012), and Radisavljevic-Gajic (2015c). It will be an interesting research topic to design an observer using the two-stage feedback design technique.

2.6 Notes

Sections 2.1, 2.2, and 2.3 are based on our journal paper Radisavljevic-Gajic and Rose (2014). Section 2.4 follows presentation of Radisavljevic-Gajic et al. (2015). Permissions for the use of such material in this research monograph were granted to us by Elsevier for the journal paper Radisavljevic-Gajic and Rose (2014) that appeared in *International Journal of Hydrogen Energy* and by the American Society of Mechanical Engineers (ASME) for the Radisavljevic-Gajic et al. (2015) ASME Dynamic Systems and Control Conference paper.

Chapter 3
Discrete-Time Two-Stage Feedback Controller Design

In this chapter the two-stage feedback controller design for linear discrete-time control systems is presented by following the results of Radisavljevic-Gajic (2015a), Sect. 3.1. The design algorithm is specialized and simplified for linear systems with slow and fast modes, known as singularly perturbed linear discrete-time systems. Since there are two formulations of discrete-time singularly perturbed systems, the results are presented for both of them as discussed in Radisavljevic-Gajic (2015a, b), Sects. 3.2 and 3.3. The conditions needed for applicability of the presented two-stage design in two time scales are established. The proposed two-stage feedback controller design procedure and its very high accuracy were demonstrated on the eigenvalue assignment problem and the mixed linear-quadratic optimal controller/eigenvalue assignment controller problem for the steam power system, hydropower system, and proton exchange membrane fuel cell.

3.1 Discrete-Time Two-Stage Feedback Controller Design

Consider a linear discrete time-invariant dynamic system represented by the following difference equation (Ogata 1995):

$$x(k+1) = \begin{bmatrix} x_1(k+1) \\ x_2(k+1) \end{bmatrix} = \begin{bmatrix} A_{11} & A_{12} \\ A_{21} & A_{22} \end{bmatrix} \begin{bmatrix} x_1(k) \\ x_2(k) \end{bmatrix} + \begin{bmatrix} B_{11} \\ B_{22} \end{bmatrix} u(k) = Ax(k) + Bu(k) \tag{3.1}$$

where $x(k) \in \mathrm{R}^n$, $x_1(k) \in \mathrm{R}^{n_1}$, and $x_2(k) \in \mathrm{R}^{n_2}$, $n = n_1 + n_2$, are state space variables, $u(k) \in \mathrm{R}^m$ is the control input vector, and A_{ij} and B_{ii}, $i, j = 1, 2$, are constant matrices of appropriate dimensions. Matrices A_{11} and A_{22} define subsystems of dimensions n_1

V. Radisavljević-Gajić et al., *Multi-Stage and Multi-Time Scale Feedback Control of Linear Systems with Applications to Fuel Cells*, Mechanical Engineering Series, https://doi.org/10.1007/978-3-030-10389-7_3

and n_2, respectively, corresponding to the state space variables $x_1(k)$ and $x_2(k)$. Matrices A_{12} and A_{21} define couplings between the subsystems.

The algorithm of Radisavljevic-Gajic (2015a) presented a two-stage linear feedback controller design for (3.1) in terms of subsystems, which can be used either for the purpose of subsystem stabilization or system eigenvalue assignment or for design of linear-quadratic optimal controllers (Sinha 2007), observers, and filters (Ogata 1995; Sinha 2007; Chen 2012). The algorithm can be used, for example, to set up the closed-loop eigenvalues of (3.1) in desired locations in two stages by performing operations on subsystems only. In general, it can be used for design of any full-state linear feedback controller. This is particularly important for linear dynamic systems with slow and fast state variables that have a natural decomposition into two subsystems, which will be considered in Sects. 3.2 and 3.3.

The two-stage feedback controller design algorithm of Radisavljevic-Gajic (2015a) is presented below into two design stages that can be broken down into five steps.

Stage 1: Second Subsystem Controller Design
Step 1: Use the following change of variables:

$$\begin{aligned}\begin{bmatrix} x_1(k) \\ \eta(k) \end{bmatrix} &= T_1 \begin{bmatrix} x_1(k) \\ x_2(k) \end{bmatrix} = \begin{bmatrix} I & 0 \\ L & I \end{bmatrix} \begin{bmatrix} x_1(k) \\ x_2(k) \end{bmatrix} \\ \begin{bmatrix} x_1(k) \\ x_2(k) \end{bmatrix} &= T_1^{-1} \begin{bmatrix} x_1(k) \\ \eta(k) \end{bmatrix} = \begin{bmatrix} I & 0 \\ -L & I \end{bmatrix} \begin{bmatrix} x_1(k) \\ \eta(k) \end{bmatrix}\end{aligned} \tag{3.2}$$

where L satisfies a nonsymmetric, nonsquare, nonlinear algebraic equation

$$LA_{11} - A_{22}L - LA_{12}L + A_{21} = 0 \tag{3.3}$$

which leads to the state space difference equation with the upper block triangular system matrix

$$\begin{aligned}\begin{bmatrix} x_1(k+1) \\ \eta(k+1) \end{bmatrix} &= \begin{bmatrix} A_{11} - A_{12}L & A_{12} \\ 0 & A_{22} + LA_{12} \end{bmatrix} \begin{bmatrix} x_1(k) \\ \eta(k) \end{bmatrix} + \begin{bmatrix} B_{11} \\ B_{22} + LB_{11} \end{bmatrix} u(k) \\ &= \begin{bmatrix} A_1 & A_{12} \\ 0 & A_2 \end{bmatrix} \begin{bmatrix} x_1(k) \\ \eta(k) \end{bmatrix} + \begin{bmatrix} B_{11} \\ B_2 \end{bmatrix} u(k) \\ A_1 = A_{11} - A_{12}L, \quad & A_2 = A_{22} + LA_{12}, \quad B_2 = B_{22} + LB_{11}\end{aligned} \tag{3.4}$$

Nonlinear matrix algebraic equation (3.3) represents a nonsymmetric nonsquare, Riccati-type, algebraic equation. It has been studied in control literature from both theoretical and numerical point of views; see, for example, Medanic (1982) and Gao and Bai (2010).

Step 2: Apply feedback control

$$u(k) = -G_2\eta(k) + v(k)$$

to set up the closed-loop eigenvalues of the η-subsystem, which leads to the following difference equation:

$$\begin{bmatrix} x_1(k+1) \\ \eta(k+1) \end{bmatrix} = \begin{bmatrix} A_1 & A_{12} - B_{11}G_2 \\ 0 & A_2 - B_2G_2 \end{bmatrix} \begin{bmatrix} x_1(k) \\ \eta(k) \end{bmatrix} + \begin{bmatrix} B_{11} \\ B_2 \end{bmatrix} v(k) \tag{3.5}$$

Stage 2: First Subsystem Controllers Design

Step 3: Use another change of state variables as

$$\begin{aligned} \begin{bmatrix} \xi(k) \\ \eta(k) \end{bmatrix} &= T_2 \begin{bmatrix} x_1(k) \\ \eta(k) \end{bmatrix} = \begin{bmatrix} I & -P \\ 0 & I \end{bmatrix} \begin{bmatrix} x_1(k) \\ \eta(k) \end{bmatrix} \\ \begin{bmatrix} x_1(k) \\ \eta(k) \end{bmatrix} &= T_2^{-1} \begin{bmatrix} \xi(k) \\ \eta(k) \end{bmatrix} = \begin{bmatrix} I & P \\ 0 & I \end{bmatrix} \begin{bmatrix} x_1(k) \\ \eta(k) \end{bmatrix} \end{aligned} \tag{3.6}$$

where P satisfies the Sylvester algebraic equation

$$A_1P - P(A_2 - B_2G_2) + A_{12} - B_{11}G_2 = 0 \tag{3.7}$$

The unique solution of this algebraic equation exists under the assumption that matrices A_1 and $A_2 - B_2G_2$ have no eigenvalues in common (Chen 2012). The second transformation produces

$$\begin{bmatrix} \xi(k+1) \\ \eta(k+1) \end{bmatrix} = \begin{bmatrix} A_1 & 0 \\ 0 & A_2 - B_2G_2 \end{bmatrix} \begin{bmatrix} \xi(k) \\ \eta(k) \end{bmatrix} + \begin{bmatrix} B_{11} - PB_2 \\ B_2 \end{bmatrix} v(k) \tag{3.8}$$

Step 4: Apply state feedback

$$v(k) = -G_1\xi(k)$$

to control the ξ-subsystem, that is,

$$\begin{aligned} \begin{bmatrix} \xi(k+1) \\ \eta(k+1) \end{bmatrix} &= \begin{bmatrix} A_1 - B_1G_1 & 0 \\ -B_2G_1 & A_2 - B_2G_2 \end{bmatrix} \begin{bmatrix} \xi(k) \\ \eta(k) \end{bmatrix} \\ B_1 &= B_{11} - PB_2 \end{aligned} \tag{3.9}$$

Matrices A_1, A_2, B_2 are defined in (3.4). Since (3.9) is a lower block triangular matrix, its eigenvalues are the union of the eigenvalues of $A_2 - B_2G_2$ and $A_1 - B_1G_1$ both set up at the subsystem levels using the presented two-stage design technique and appropriate feedback gains G_1 and G_2.

Step 5: The feedback gains can be mapped into the original coordinates using the corresponding transformation, which leads to

$$u(x(k)) = -[G_1 + (G_2 - G_1P)L \;\; G_2 - G_1P]\begin{bmatrix} x_1(k) \\ x_2(k) \end{bmatrix} \tag{3.10}$$

$$= -[\,G_{1eq} \quad G_{2eq}\,]x(k) = -G_{eq}x(t)$$

$$G_{1eq} = G_1 + (G_2 - G_1P)L, \quad G_{2eq} = G_2 - G_1P$$

In summary, the design of Radisavljevic-Gajic (2015a) requires a solution of the nonlinear algebraic equation (3.3) for L of order $n_1 \times n_2$ and the Sylvester algebraic equation (3.7) of order $n_1 \times n_2$ for P. Since the design is done in two independent stages, inaccuracies made in the second stage will not affect the first stage design accuracy.

For the eigenvalues assignment problem, the corresponding subsystems must be controllable (Ogata 1995; Sinha 2007; Chen 2012). Consequently, Steps 2 and 4 are applicable under the following assumptions.

Assumption 3.1 The pair (A_1, B_1) is controllable.

Assumption 3.2 The pair (A_2, B_2) is controllable.

For the stabilization problem only, Assumptions 3.1 and 3.2 can be relaxed to the stabilizability requirements imposed on the pairs (A_1, B_1) and (A_2, B_2). In the case that one intends to design linear-quadratic optimal controllers for subsystems, Assumptions 3.1 and 3.2 should be replaced by the stabilizability conditions of the corresponding pairs and augmented by the detectability conditions imposed on the corresponding subsystem and performance criterion penalty matrices (Sinha 2007).

In addition, the existence of a unique solution of the Sylvester algebraic equation require the next assumption.

Assumption 3.3 The unique solution of the Sylvester algebraic equation (3.7) exists under the assumption that matrices A_1 and $A_2 - B_2G_2$ have no eigenvalues in common (Chen 2012).

Comment 3.1 Note that the matrices given in Assumptions 3.1–3.3 are parameterized by the solution matrices of the L- and P-equations, that is, $A_1 = A_1(L)$, $A_2 = A_2(L)$, $B_1 = B_1(L, P)$, $B_2 = B_2(L)$, so that for certain choices of L and P matrices, the assumptions will be either satisfied or not satisfied. This gives an extra degree of freedom in choosing matrix L. Having found L, the matrix P is uniquely determined from (3.7) under Assumption 3.3 (it is also parameterized by L), that is, $P = P(L)$. Since the matrix L is obtained via the eigenvector approach, one may study the impact on the open-loop transient response (and other open-loop system quantities) of the eigenspace spanned by the eigenvectors used to generate the solution matrix L. However, since the two-stage feedback algorithm will be used, the system closed-loop essential quantities will be determined by the closed-loop feedback matrices $A_1 - B_1G_1$ and $A_2 - B_2G_2$.

The reduction in computational requirements comes first of all from the fact that instead of designing a full-state feedback controller for the entire system of order $n \times n$, that is, by finding the matrix G such that the closed-loop system matrix $A - BG$ has desired characteristics, the design is done at the subsystem levels, that is, by finding the subsystem full-state feedback gains $G_1$and G_2 such that the closed-loop subsystems defined by matrices $A_1 - B_1G_1$ and $A_2 - B_2G_2$ (of dimensions $n_1 \times n_1$ and $n_2 \times n_2$, respectively, $n = n_1+n_2$) have desired characteristics. Depending on the complexity and type of the required full-state design: eigenvalue placement, optimal control in some sense, robust control, observer-based controller, Kalman filter-based controller for linear discrete-time stochastic systems, and so on, the reduction in required computations can be substantial.

3.2 Slow-Fast Design for Systems Defined in the Slow Time Scale

In the discrete-time domain, there are two formulations of discrete-time singularly perturbed linear systems: the *slow-time-scale formulation*, introduced in Rao and Naidu (1981) and Naidu and Calise (2001), and the *fast-time-scale formulation* originally derived in Litkouhi and Khalil (1984, 1985) and further developed in Bidani et al. (2002) and Kim et al. (2004). In this section, we will show how these results can be used to study the slow-time-scale formulation of Rao and Naidu (1981) and Naidu and Calise (2001) of singularly perturbed linear discrete-time systems. The fast-time-scale formulation of singularly perturbed discrete-time systems and the corresponding two-stage feedback controller design will be considered in Sect. 3.3.

The time-invariant linear discrete-time singularly perturbed system is represented using the slow-time-scale formulation by the following difference equation

$$\begin{bmatrix} x_1(k+1) \\ x_2(k+1) \end{bmatrix} = \begin{bmatrix} A_{11} & \varepsilon A_{12} \\ A_{21} & \varepsilon A_{22} \end{bmatrix} \begin{bmatrix} x_1(k) \\ x_2(k) \end{bmatrix} + \begin{bmatrix} B_{11} \\ B_{22} \end{bmatrix} u(k) \tag{3.11}$$

where ε is a small positive singular perturbation parameter that indicates separation of state space variables into slow ones, $x_1(k)$, and the fast state variables $x_2(k)$. The dimensions of state space variables, system control input, and constant matrices in (3.11) are defined in Sect. 3.1 after equation (3.1). The value of the small positive singular perturbation parameter can be approximately taken as a ratio of the magnitudes of the slowest fast eigenvalue and the fastest slow eigenvalue (Kokotovic et al. 1999; Naidu and Calise 2001), that is,

$$\varepsilon \approx \frac{\max\{|\lambda_f(A)|\}}{\min\{|\lambda_s(A)|\}} \tag{3.12}$$

It is a standard assumption in theory of singularly perturbed linear discrete-time systems represented in the fast-time-scale formulation that matrix A_{11} is nonsingular (Naidu and Calise 2001). Hence, the following assumption is imposed.

Assumption 3.4 The subsystem matrix A_{11} is nonsingular.

The two-stage feedback design from Sect. 3.1 applied to the singularly perturbed linear discrete-time system defined in (3.11) has the following five steps.

Stage 1: Fast Subsystem Feedback Controller Design

Step 1: Solve the algebraic equation (3.3), which in this case has the form

$$LA_{11} - \varepsilon A_{22}L - \varepsilon LA_{12}L + A_{21} = 0 \tag{3.13}$$

Since the nonlinear term for L in (3.13) is multiplied by the small singular perturbation parameter ε, solving (3.13) is much easier than solving (3.3). As a matter of fact, two algorithms (fixed-point iterations and Newton algorithm) can be derived in terms of *linear algebraic equations* for solving (3.3). In both cases a unique solution of (3.13) exists for sufficiently small values of ε under Assumption 3.4. Note that by setting $\varepsilon = 0$ in (3.13), the zero-order approximation for L is given by

$$L^{(0)} = -A_{21}A_{11}^{-1}, \quad L = L^{(0)} + O(\varepsilon) \tag{3.14}$$

The fixed-point algorithm is obtained in a simple manner by delaying by one iteration all terms multiplied by ε. By performing fixed-point iterations on (3.13), we are faced with the problem of solving a system of linear algebraic equations in each iteration defined by

$$L^{(i+1)}A_{11} = -A_{21} + \varepsilon\big(A_{22} + L^{(i)}A_{12}\big)L^{(i)}, \quad L^{(0)} = -A_{21}A_{11}^{-1}, \quad i = 1,2, \ldots \tag{3.15}$$

This algorithm converges with the linear rate of convergence of the order of ε, denoted by $O(\varepsilon)$, which means that after i iterations, the accuracy of $O(\varepsilon^i)$[1] is achieved, that is, the accuracy is improved in each iteration by $O(\varepsilon)$. The convergence is achieved if the radius of convergence of the fixed-point iterative scheme (3.15) is smaller than one in each iteration.

Another alternative to solve (3.13) for small values ε is to use the Newton method for solving weakly nonlinear algebraic equation (3.13). It is well-known that the Newton method has a quadratic rate of convergence, assuming that the chosen initial

[1] $O(\varepsilon^i)$ is defined as $O(\varepsilon^i) < c\varepsilon^i$, where c is a bounded constant and i is a real number.

guess is very good, which is true in this case since $L^{(0)} = L + O(\varepsilon)$. The application of the Newton method leads to iterations involving linear algebraic equations as

$$\left(A_{22} - \varepsilon L^{(i)} A_{12}\right) L^{(i+1)} + \varepsilon L^{(i+1)} \left(A_{11} - L^{(i)} A_{12}\right) = A_{21} - \varepsilon L^{(i)} A_{12} L^{(i)},$$
$$L^{(0)} = -A_{21} A_{11}^{-1}, \quad i = 1,2, \ldots \tag{3.16}$$

If ε is not sufficiently small, neither the fixed-point iterations nor the Newton method will converge, and the eigenvector method discussed in Sect. 2.1 (see formulas (2.17), (2.18), and (2.19)) needs to be used to solve nonlinear algebraic equation (3.13).

Comment 3.2 It follows from (3.13) and (3.14) and Assumption 3.4 that the solution of (3.13) is unique for sufficiently small values of ε (implicit function theorem, Ortega and Reinhardt 2000). Moreover, according to (3.14) that solution is $O(1)$.

Now we use the following change of variables:

$$\begin{aligned} \begin{bmatrix} x_1(k) \\ x_f(k) \end{bmatrix} &= T_{1sp} \begin{bmatrix} x_1(k) \\ x_2(k) \end{bmatrix} = \begin{bmatrix} I & 0 \\ L & I \end{bmatrix} \begin{bmatrix} x_1(k) \\ x_2(k) \end{bmatrix} \\ \begin{bmatrix} x_1(k) \\ x_2(k) \end{bmatrix} &= T_{1sp}^{-1} \begin{bmatrix} x_1(k) \\ x_f(k) \end{bmatrix} = \begin{bmatrix} I & 0 \\ -L & I \end{bmatrix} \begin{bmatrix} x_1(k) \\ x_f(k) \end{bmatrix} \end{aligned} \tag{3.17}$$

which transforms the discrete-time system (3.11) into

$$\begin{aligned} \begin{bmatrix} x_1(k+1) \\ x_f(k+1) \end{bmatrix} &= \begin{bmatrix} A_{11} - \varepsilon A_{12} L & \varepsilon A_{12} \\ 0 & \varepsilon(A_{22} + L A_{12}) \end{bmatrix} \begin{bmatrix} x_1(k) \\ x_f(k) \end{bmatrix} + \begin{bmatrix} B_{11} \\ B_{22} + L B_{11} \end{bmatrix} u(k) \\ &= \begin{bmatrix} A_s & \varepsilon A_{12} \\ 0 & \varepsilon A_f \end{bmatrix} \begin{bmatrix} x_1(k) \\ x_f(k) \end{bmatrix} + \begin{bmatrix} B_{11} \\ B_f \end{bmatrix} u(k) \\ A_s = A_{11} - \varepsilon A_{12} L, &\quad A_f = A_{22} + L A_{12}, \quad B_f = B_{22} + L B_{11} \end{aligned} \tag{3.18}$$

Step 2: Apply feedback control $u(k) = -\varepsilon G_f\, x_f\,(k) + v(k)$ to set up the closed-loop eigenvalues of the *fast subsystem*, which leads to

$$\begin{bmatrix} x_1(k+1) \\ x_f(k+1) \end{bmatrix} = \begin{bmatrix} A_s & \varepsilon\left(A_{12} - B_{11} G_f\right) \\ 0 & \varepsilon\left(A_f - B_f G_f\right) \end{bmatrix} \begin{bmatrix} x_1(k) \\ x_f(k) \end{bmatrix} + \begin{bmatrix} B_{11} \\ B_f \end{bmatrix} v(k) \tag{3.19}$$

It should be emphasized that the proper scaling of the fast subsystem feedback gain by ε is essential for the completion of the two-stage slow-fast feedback controller design.

Stage 2: Slow Subsystem Feedback Controller Design
Step 3: Use another change of state variables as

$$\begin{bmatrix} x_s(k) \\ x_f(k) \end{bmatrix} = T_{2sp}\begin{bmatrix} x_1(k) \\ x_f(k) \end{bmatrix} = \begin{bmatrix} I & -\varepsilon P \\ 0 & I \end{bmatrix}\begin{bmatrix} x_1(k) \\ x_f(k) \end{bmatrix}$$
$$\begin{bmatrix} x_1(k) \\ x_f(k) \end{bmatrix} = T_{2sp}^{-1}\begin{bmatrix} x_s(k) \\ x_f(k) \end{bmatrix} = \begin{bmatrix} I & \varepsilon P \\ 0 & I \end{bmatrix}\begin{bmatrix} x_s(k) \\ x_f(k) \end{bmatrix} \tag{3.20}$$

where P satisfies the Sylvester algebraic equation

$$A_sP - \varepsilon P(A_f - B_fG_f) + A_{12} - B_{11}G_f = 0 \tag{3.21}$$

with matrices A_s, A_f, B_f defined in (3.18). Having obtained the solution matrix L from (3.13), matrices A_s, A_f, B_f can be calculated from (3.18), with matrix G_f being determined in Step 2 to control the fast subsystem (e.g., to set up its closed-loop eigenvalues in the desired locations). The algebraic equation (3.21) for P can be solved directly as a linear algebraic Sylvester equation. A unique solution of the Sylvester algebraic equation exists if matrices A_s and $\varepsilon(A_f - B_f G_f)$ have no eigenvalues in common (Chen 2012), that is, if the following relationship holds $\varepsilon\lambda(A_f - B_f G_f) = O(\varepsilon) \neq \lambda(A_s) = O(1)$, which is always satisfied since the slow and fast subsystem eigenvalues are of the different order of magnitude. Algebraic equation (3.21) can be also solved by running fixed-point iterations on a system of linear algebraic equations as follows

$$A_sP^{(i+1)} - \varepsilon P^{(i)}(A_f - B_fG_f) + A_{12} - B_{11}G_f = 0$$
$$P^{(0)} = -A_s^{-1}(A_{12} - B_{11}G_f), \quad i = 0,1,2, \ldots \tag{3.22}$$

which has the rate of convergence $O(\varepsilon)$, meaning that after i iterations, the accuracy of $O(\varepsilon^i)$ is achieved.

The state transformation (3.20) leads to

$$\begin{bmatrix} x_s(k+1) \\ x_f(k+1) \end{bmatrix} = \begin{bmatrix} A_s & 0 \\ 0 & \varepsilon(A_f - B_fG_f) \end{bmatrix}\begin{bmatrix} x_s(k) \\ x_f(k) \end{bmatrix} + \begin{bmatrix} B_s \\ B_f \end{bmatrix} v(k)$$
$$B_s = B_{11} - \varepsilon PB_f \tag{3.23}$$

Step 4: Apply state feedback

$$v(k) = -G_sx_s(k)$$

to control the slow subsystem, which produces

$$\begin{bmatrix} x_s(k+1) \\ x_f(k+1) \end{bmatrix} = \begin{bmatrix} A_s - B_sG_s & 0 \\ -B_fG_s & \varepsilon(A_f - B_fG_f) \end{bmatrix}\begin{bmatrix} x_s(k) \\ x_f(k) \end{bmatrix} \tag{3.24}$$

Since (3.24) is a lower block triangular matrix, its eigenvalues are the union of the eigenvalues of $\varepsilon(A_f - B_f G_f)$ and the eigenvalues of matrix $A_s - B_s G_s$ both set up at the subsystem levels. Note that the closed-loop *slow* subsystem matrix eigenvalues are $O(1)$, placed close to the unit circle, and the closed-loop *fast* eigenvalues are in the ε-neighborhood of the origin, that is, they are $O(\varepsilon)$ and hence change fast as $O(\varepsilon^k)$, $k = 1,2,3,\ldots$.

Step 5: The feedback gains in the original coordinates are obtained using the following steps

$$\begin{aligned}
u(x(k)) &= -[0 \quad \varepsilon G_f]\begin{bmatrix} x_1(k) \\ x_f(k) \end{bmatrix} - [G_s \quad 0]\begin{bmatrix} x_s(k) \\ x_f(k) \end{bmatrix} \\
&= -[0 \quad \varepsilon G_f]\begin{bmatrix} x_1(k) \\ x_f(k) \end{bmatrix} - [G_s \quad 0]T_{2sp}\begin{bmatrix} x_1(k) \\ x_f(k) \end{bmatrix} \\
&= -\{[0 \quad \varepsilon G_f] + [G_s \quad 0]T_{2sp}\}T_{1sp}\begin{bmatrix} x_1(k) \\ x_2(k) \end{bmatrix} \\
&= -\left\{[0 \quad \varepsilon G_f] + [G_s \quad 0]\begin{bmatrix} I & -\varepsilon P \\ 0 & I \end{bmatrix}\right\}\begin{bmatrix} I & 0 \\ L & I \end{bmatrix}\begin{bmatrix} x_1(k) \\ x_2(k) \end{bmatrix} \\
&= -[G_s \quad \varepsilon G_f - \varepsilon G_s P]\begin{bmatrix} I & 0 \\ L & I \end{bmatrix}\begin{bmatrix} x_1(k) \\ x_2(k) \end{bmatrix} \\
&= -\left[G_s + \varepsilon\big(G_f - G_s P\big)L \quad \varepsilon\big(G_f - G_s P\big)\right]\begin{bmatrix} x_1(k) \\ x_2(k) \end{bmatrix} \\
&= -G_{1eq}x_1(k) - G_{2eq}x_2(k) = -G_{eq}x(k)
\end{aligned} \tag{3.25}$$

with the equivalent feedback gains in the original coordinates from the original state space variables given by

$$G_{1eq} = G_s + \varepsilon\big(G_f - G_s P\big)L = G_s + G_{2eq}L, \quad G_{2eq} = \varepsilon\big(G_f - G_s P\big) \tag{3.26}$$

In summary, to perform the linear feedback design in two stages for singularly perturbed linear discrete-time systems, we need to solve algebraic equations (3.13) and (3.21), which have, respectively, simpler forms than the corresponding nonlinear algebraic equation (3.3) and linear algebraic equation (3.7) of the general two-stage feedback design algorithm.

Comment 3.3 As defined, respectively, by equations (3.13) and (3.21) and the corresponding numerical algorithms (3.15) or (3.16) and (3.22) used for solving these equations, the solution matrices L and P are both $O(1)$. As such, they will not interfere with the system slow-fast dynamics, and the slow state variables will remain slow, and the fast state variables will remain fast after two similarity transformations (3.17) and (3.20) are applied.

3.2.1 Example: A Power System

In this example, we perform the eigenvalue assignment design in two stages of a power system by calculating independently slow and fast subsystem feedback gains using the reduced-order dynamic models and corresponding reduced-order matrices. This system has four slow ($n_1 = 4$) and four fast ($n_2 = 4$) state variables (modes). The small singular perturbation parameter is taken as $\varepsilon = 0.259$. It represents roughly the ratio of the magnitude of the slowest fast eigenvalue and the magnitude of the fastest slow eigenvalue. The problem matrices for the state space model of (3.1), that is, $x(k+1) = Ax(k)+Bu(k)$, are given by

$$A = \begin{bmatrix} 0.835 & 0 & 0 & 0 & 0 & 0 & 0 & 0 \\ 0.096 & 0.861 & 0 & 0 & 0 & 0 & 0 & 0 \\ -0.002 & -0.005 & 0.882 & -0.253 & 0.041 & -0.003 & -0.025 & -0.001 \\ 0.007 & 0.014 & -0.029 & 0.928 & 0 & 0.006 & 0.0059 & 0.002 \\ -0.030 & -0.061 & 2.028 & -2.303 & -0.088 & -0.021 & -0.224 & -0.008 \\ 0.048 & 0.758 & 0 & 0 & 0 & 0.165 & 0 & 0.023 \\ -0.012 & -0.027 & 1.209 & -1.400 & 0.161 & -0.013 & 0.156 & 0.006 \\ 0.815 & 0 & 0 & 0 & 0 & 0 & 0 & 0.011 \end{bmatrix}$$

$$B^T = \begin{bmatrix} 0 & 0 & 0.294 & -0.038 & 2.762 & 0 & 1.473 & 0 \\ 3.295 & 0.152 & -0.003 & 0.010 & -0.051 & 0.056 & -0.015 & 2.477 \end{bmatrix}$$

The eigenvalues of matrix A are

$$\lambda(A) = \{0.0110 \quad 0.0184 \quad 0.1650 \quad 0.2866 \quad 0.8350 \quad 0.8910 \quad 0.8745 \pm j0.1696\}$$

The first four eigenvalues close to the origin are the fast eigenvalues, and the remaining four eigenvalues close to the unit circle are the slow eigenvalues.

Consistently to the partitioned singularly perturbed system structure defined in (3.11), the blocks (1,2) and (2,2) of matrix A (each of dimension 4x4) are multiplied by ε. Consequently, the block A_{12} in (3.11) is the block (1,2) of A divided by ε, and the block A_{22} in (3.11) is the block (2,2) of A divided by ε. Using $\varepsilon = 0.259$, the corresponding subsystem matrices consistent with (3.11) are given as

$$A_{11}=\begin{bmatrix}0.835 & 0 & 0 & 0\\ 0.096 & 0.861 & 0 & 0\\ -0.002 & -0.005 & 0.882 & -0.253\\ 0.007 & 0.014 & -0.029 & 0.928\end{bmatrix},\quad A_{12}=\begin{bmatrix}0 & 0 & 0 & 0\\ 0 & 0 & 0 & 0.112\\ 0.158 & -0.012 & -0.097 & -0.004\\ 0 & 0.023 & 0.228 & 0.008\end{bmatrix}$$

$$A_{21}=\begin{bmatrix}-0.030 & -0.061 & 2.028 & -2.303\\ 0.048 & 0.758 & 0 & 0\\ -0.012 & -0.027 & 1.209 & -1.400\\ 0.815 & 0 & 0 & 0\end{bmatrix},\quad A_{22}=\begin{bmatrix}0.340 & -0.081 & -0.865 & -0.031\\ 0 & 0.037 & 0 & 0.089\\ 0.622 & -0.050 & 0.602 & 0.023\\ 0 & 0 & 0 & 0.043\end{bmatrix}$$

$$B_{11}=\begin{bmatrix}0 & 3.295\\ 0 & 0.152\\ 0.294 & -0.003\\ -0.038 & 0.001\end{bmatrix},\quad B_{22}=\begin{bmatrix}2.762 & -0.051\\ 0 & 0.056\\ 1.473 & -0.015\\ 0 & 2.477\end{bmatrix}$$

It can be easily checked that the eigenvalues of A_{11} are small perturbations of the slow eigenvalues and the eigenvalues of εA_{22} are small perturbation of the fast eigenvalues, which indicates that the partitioning done is consistent with the slow-fast nature of this dynamic system. Sometimes, it is more difficult to separate the slow and the fast state variables. In such a case, we have to use permutation of state variables and exchange some rows and columns in the corresponding state space model matrix in order to get the singularly perturbed structure defined in (3.11).

The solution of L-equation (3.13) is obtained by using the fixed-point algorithm (3.15), in which case we run 30 iterations on the corresponding linear algebraic equations to obtain accuracy of $E^{(i)}=O(10^{-14})$, where the accuracy is defined as

$$\| E^{(i)} \| = \| L^{(i)}A_{11} - \varepsilon A_{22}L^{(i)} - \varepsilon L^{(i)}A_{12}L^{(i)} + A_{21} \|$$

Note that this accuracy is consistent with the system slow-fast separation dynamics since we have $(0.2866/0.8350)^{30}=1.1687\times 10^{-14}$. Having obtained $L^{(30)}\approx L$, we calculated A_s, B_s, A_f, B_f using (3.18) and (3.23), which produced

$$L\approx L^{(30)}=\begin{bmatrix}0.0148 & 0.0337 & -2.0799 & 1.8132\\ 0.0887 & -1.0441 & 0 & 0\\ -0.0221 & -0.0108 & -2.2552 & 1.5242\\ -0.9891 & 0 & 0 & 0\end{bmatrix}\Rightarrow$$

$$A_s=\begin{bmatrix}0.8353 & 0 & 0 & 0\\ 0.1247 & 0.8910 & 0 & 0\\ -0.0039 & -0.0098 & 0.9109 & -0.2892\\ 0.0097 & 0.0209 & 0.1041 & 0.8381\end{bmatrix},\quad A_f=\begin{bmatrix}0.0105 & -0.0150 & -0.2511 & -0.0051\\ 0 & 0.6371 & 0 & -0.0281\\ 0.2646 & 0.0112 & 1.1672 & 0.0424\\ 0 & 0 & 0 & 0.0425\end{bmatrix},$$

$$B_s = \begin{bmatrix} 0.8476 & 4.0776 \\ -0.1294 & 0.0159 \\ 0.3919 & 0.0054 \\ 0.0433 & -0.0054 \end{bmatrix}, \quad B_f = \begin{bmatrix} 2.0816 & 0.0272 \\ 0 & 0.1896 \\ 0.7520 & -0.0673 \\ 0 & -0.7820 \end{bmatrix}$$

Assume that we want to assign the closed-loop slow eigenvalues at $\lambda_s^{desired} = \{0.9 \pm j0.005,\ 0.85,\ 0.80\}$ and the closed-loop fast eigenvalues at $\lambda_f^{desired} = \{0.2 \pm j0.01,\ 0.15,\ 0.10\}$. According to the presented two-stage feedback design, we first find the fast subsystem gain G_f (Step 2) that locates $\lambda(A_f - B_f G_f) = \lambda_{fast}^{desired}$. This leads to the following result (using the MATLAB function "place")

$$G_f = \begin{bmatrix} -0.0777 & -0.0071 & 0.1441 & 0.0553 \\ -0.2249 & -0.0751 & -0.1652 & 0.7543 \end{bmatrix}$$

In Step 3, we solve the Sylvester algebraic equation (3.21) directly using the MATLAB function "lyap," which produced

$$P = \begin{bmatrix} -1.2535 & -0.4624 & -0.8820 & 3.7841 \\ 0.1932 & 0.0785 & 0.1293 & -0.6572 \\ -0.2073 & -0.0010 & 0.0715 & 0.0278 \\ -0.0010 & -0.0316 & -0.4146 & -0.0481 \end{bmatrix}$$

In Step 4, we find the slow subsystem feedback gain using $\lambda(A_s - B_s G_s) = \lambda_{slow}^{desired}$ and the corresponding MATLAB function "place," which produced

$$G_s = \begin{bmatrix} 0.2146 & 0.0839 & 0.3895 & -0.8643 \\ -0.0636 & -0.0253 & -0.1132 & 0.1743 \end{bmatrix}$$

It is important to emphasize that the slow and fast gains G_s, G_f are obtained using calculations with the reduced-order matrices. The actual full-state feedback gain applied to the system is obtained in Step 5 as

$$\begin{aligned} G_{eq} &= [G_{1eq} \;\; G_{2eq}] = \left[G_s + G_{2eq}L \;\; \varepsilon\left(G_f - G_s P\right)\right] \\ &= \begin{bmatrix} 0.4104 & 0.0705 & 0.2893 & -0.7697 & 0.0661 & 0.0152 & -0.0165 & -0.1953 \\ -0.3199 & -0.0015 & 0.1412 & -0.0317 & -0.0837 & -0.0251 & -0.0356 & 0.2564 \end{bmatrix} \end{aligned}$$

It can be easily check (using MATLAB) that indeed this gain produces the desired set of the closed-loop eigenvalues with the accuracy of $O(10^{-14})$, namely, we have

$$\begin{aligned} \lambda\left(A - BG_{eq}\right) &= \lambda_{system}^{desired} = \lambda_{slow}^{desired} \cup \lambda_{fast}^{desired} \\ &= \{0.90000000000000 \pm j0.00500000000000 \quad 0.85000000000000 \quad 0.80000000000000\} \\ &\cup \{0.20000000000000 \pm j0.01000000000000 \quad 0.15000000000000 \quad 0.10000000000000\} \end{aligned}$$

If we use only the zeroth-order approximation for the corresponding feedback design ($i = 0$), which was done in almost all papers dealing with the discrete-time singularly perturbed systems (Singh et al. 1996; Wang et al. 1996; Naidu and Calise, 2001), we get a poor approximation for the closed-loop desired eigenvalues given by

$$\lambda_{system}^{(0)} = \{0.8406 \pm j0.0097 \quad 1.0511 \quad 0.7962 \quad 0.1596 \pm j0.1350 \quad 0.1600 \quad 0.0925\}$$

Since in the proposed method the accuracy can be simply increased just by running additional iterations, we get easily $\lambda_{system}^{(1)}, \lambda_{system}^{(2)}, \lambda_{system}^{(3)}, \ldots$. For example, for $i = 1$, we have

$$\lambda_{system}^{(1)} = \{0.9010 \pm j0.0051 \quad 0.8384 \quad 0.7932 \quad 0.2125 \pm j0.0054 \quad 0.1460 \quad 0.0954\}$$

and for $i = 7$, the accuracy of $O(10^{-4})$ is achieved, that is,

$$\lambda_{system}^{(7)} = \{0.9000 \pm j0.0050 \quad 0.8500 \quad 0.8000 \quad 0.2000 \pm j0.0100 \quad 0.1500 \quad 0.1000\}$$

As indicated before, in this example, $i = 30$ iterations are needed to achieve the accuracy of $O(10^{-14})$.

The power of the two-stage design lies in the fact that different types of controllers can be designed for different subsystems using the corresponding feedback gains obtained by doing calculations only with the subsystem (reduced-order) matrices. For example, for the slow subsystem, we can use the linear-quadratic optimal controller, and for the fast subsystem, we can design a controller based on the eigenvalue assignment technique (and the other way around) or any other technique for design of linear feedback controllers. This introduces the concept of partial optimality on this design procedure.

3.3 Slow-Fast Design for Systems Defined in the Fast Time Scale

The corresponding time-invariant discrete-time linear singularly perturbed system represented in the fast-time-scale formulation (Litkouhi and Khalil 1984, 1985; Bidani et al. 2002; Kim et al. 2004) is defined by the following state space form:

$$\begin{bmatrix} x_1(k+1) \\ x_2(k+1) \end{bmatrix} = \begin{bmatrix} I + \varepsilon A_{11} & \varepsilon A_{12} \\ A_{21} & A_{22} \end{bmatrix} \begin{bmatrix} x_1(k) \\ x_2(k) \end{bmatrix} + \begin{bmatrix} \varepsilon B_{11} \\ B_{22} \end{bmatrix} u(k) \tag{3.27}$$

where ε is a small positive singular perturbation parameter that indicates separation of state space variables into slow ones, $x_1(k)$, and the fast state variables $x_2(k)$. The dimensions of state variables, control input, and constant matrices are defined in Sect. 3.1 in formula (3.1).

Comment 3.4 In (3.27), the identity matrix is a consequence of the discretization procedure applied to the corresponding continuous-time singularly perturbed system. The actual slow subsystem matrix is $I+\varepsilon A_{11}$ whose eigenvalues $1+\lambda(\varepsilon A_{11}) = 1+\varepsilon\lambda(A_{11})$ are close to the unit circle, and hence they represent the slow dynamics. In contrast, the fast dynamics is represented by the eigenvalues of the fast subsystem matrix A_{22} that are located inside the unit circle closer to the origin.

It is a standard assumption in theory of singularly perturbed linear discrete-time systems represented in the fast-time-scale formulation that matrix $I - A_{22}$ is nonsingular (Litkouhi and Khalil 1984, 1985; Bidani et al. 2002; Kim et al. 2004). Hence, the following assumption needs to be imposed for the derivations to be presented in the remaining part of this section.

Assumption 3.5 Fast subsystem matrix $I - A_{22}$ is nonsingular.

In the case of singularly perturbed discrete-time linear systems represented in the fast-time-scale formulation, the corresponding simplified two-stage feedback design has the following steps.

Stage 1: Control of the Fast Subsystem

Step 1: Solve the algebraic Riccati-type equation (3.3), which in the case of discrete-time singularly perturbed linear, time-invariant, dynamic systems in the fast-time-scale formulation (due to its structure defined in (3.27)) has the following form:

$$L(I+\varepsilon A_{11}) - A_{22}L - \varepsilon LA_{12}L + A_{21} = 0 \tag{3.28}$$

Apply the following change of the state space variables to the discrete-time system defined in (3.27)

$$x_f(k) = Lx_1(k) + x_2(k) \tag{3.29}$$

which leads to

$$\begin{aligned}\begin{bmatrix} x_1(k+1) \\ x_f(k+1) \end{bmatrix} &= \begin{bmatrix} I+\varepsilon(A_{11}-A_{12}L) & \varepsilon A_{12} \\ 0 & A_{22}+\varepsilon LA_{12} \end{bmatrix} \begin{bmatrix} x_1(k) \\ x_f(k) \end{bmatrix} + \begin{bmatrix} \varepsilon B_{11} \\ B_{22}+\varepsilon LB_{11} \end{bmatrix} u(k) \\ &= \begin{bmatrix} I+\varepsilon A_s & \varepsilon A_{12} \\ 0 & A_f \end{bmatrix} \begin{bmatrix} x_1(k) \\ x_f(k) \end{bmatrix} + \begin{bmatrix} \varepsilon B_{11} \\ B_f \end{bmatrix} u(k)\end{aligned} \tag{3.30}$$

where

$$A_s = A_{11} - A_{12}L, \quad A_f = A_{22} + \varepsilon LA_{12}, \quad B_f = B_{22} + \varepsilon LB_{11} \tag{3.31}$$

The original and new state variables are related via a similarity transformation given by

$$\begin{aligned}\begin{bmatrix} x_1(k) \\ x_f(k) \end{bmatrix} &= T_{1sp}\begin{bmatrix} x_1(k) \\ x_2(k) \end{bmatrix} = \begin{bmatrix} I & 0 \\ L & I \end{bmatrix}\begin{bmatrix} x_1(k) \\ x_2(k) \end{bmatrix} \\ \begin{bmatrix} x_1(k) \\ x_2(k) \end{bmatrix} &= T_{1sp}^{-1}\begin{bmatrix} x_1(k) \\ x_f(k) \end{bmatrix} = \begin{bmatrix} I & 0 \\ -L & I \end{bmatrix}\begin{bmatrix} x_1(k) \\ x_f(k) \end{bmatrix}\end{aligned} \tag{3.32}$$

Step 2: Apply feedback control

$$u(k) = -G_f x_f(k) + v(k)$$

to independently control the decoupled fast subsystem (3.30) (e.g., to set up the closed-loop eigenvalues), which leads to the following discrete-time linear system:

$$\begin{bmatrix} x_1(k+1) \\ x_f(k+1) \end{bmatrix} = \begin{bmatrix} I + \varepsilon A_s & \varepsilon(A_{12} \quad B_{11}G_f) \\ 0 & A_f - B_f G_f \end{bmatrix}\begin{bmatrix} x_1(k) \\ x_f(k) \end{bmatrix} + \begin{bmatrix} \varepsilon B_{11} \\ B_f \end{bmatrix} v(k) \tag{3.33}$$

Stage 2: Control of the Slow Subsystem

Step 3: Apply another change of state variables as

$$x_s(k) = x_1(k) - \varepsilon P x_f(k) \tag{3.34}$$

where P satisfies the Sylvester algebraic equation

$$(I + \varepsilon A_s)P - P(A_f - B_f G_f) + (A_{12} - B_{11}G_f) = 0 \tag{3.35}$$

with matrices A_s, A_f, B_f defined in (3.31). This defines another similarity transformation

$$\begin{aligned}\begin{bmatrix} x_s(k) \\ x_f(k) \end{bmatrix} &= T_{2sp}\begin{bmatrix} x_1(k) \\ x_f(k) \end{bmatrix} = \begin{bmatrix} I & -\varepsilon P \\ 0 & I \end{bmatrix}\begin{bmatrix} x_1(k) \\ x_f(k) \end{bmatrix} \\ \begin{bmatrix} x_1(k) \\ x_f(k) \end{bmatrix} &= T_{2sp}^{-1}\begin{bmatrix} x_s(k) \\ x_f(k) \end{bmatrix} = \begin{bmatrix} I & \varepsilon P \\ 0 & I \end{bmatrix}\begin{bmatrix} x_s(k) \\ x_f(k) \end{bmatrix}\end{aligned} \tag{3.36}$$

which leads to the linear discrete-time system represented by the following difference equation

$$\begin{bmatrix} x_s(k+1) \\ x_f(k+1) \end{bmatrix} = \begin{bmatrix} I + \varepsilon A_s & 0 \\ 0 & A_f - B_f G_f \end{bmatrix}\begin{bmatrix} x_s(k) \\ x_f(k) \end{bmatrix} + \begin{bmatrix} \varepsilon(B_{11} - PB_f) \\ B_f \end{bmatrix} v(k) \tag{3.37}$$

Step 4: Use state feedback

$$v(k) = -G_s x_s(k)$$

to control the slow subsystem $x_s(k+1) = (I + \varepsilon A_s)x_s(k) + \varepsilon(B_{11} - PB_f)u(k)$. Note that the slow subsystem eigenvalues are located close to the unit circle. This produces

$$\begin{bmatrix} x_s(k+1) \\ x_f(k+1) \end{bmatrix} = \begin{bmatrix} I + \varepsilon\big(A_s - \big(B_{11} - PB_f\big)G_s\big) & 0 \\ -B_f G_s & A_f - B_f G_f \end{bmatrix} \begin{bmatrix} x_s(k) \\ x_f(k) \end{bmatrix} \tag{3.38}$$

Since the system state matrix in (3.38) is a lower block triangular matrix, the system closed-loop eigenvalues are the union of the eigenvalues of the block diagonal matrices $I+\varepsilon(A_s - (B_{11} - PB_f)G_s) = I+\varepsilon[(A_{11} - A_{12}L) - (B_{11} - PB_f)G_s]$, $A_f - B_f G_f$ (Golub and Van Loan 2012). Note that both the slow and fast closed-loop eigenvalues are set up at the subsystem levels.

Step 5: The feedback gains in the original coordinates are obtained using the introduced similarity transformations

$$\begin{aligned} u(x(k)) &= -[0 \quad G_f]\begin{bmatrix} x_1(k) \\ x_f(k) \end{bmatrix} - [G_s \quad 0]\begin{bmatrix} x_s(k) \\ x_f(k) \end{bmatrix} \\ &= -[0 \quad G_f]\begin{bmatrix} x_1(k) \\ x_f(k) \end{bmatrix} - [G_s \quad 0]T_{2sp}\begin{bmatrix} x_1(k) \\ x_f(k) \end{bmatrix} \\ &= -\{[0 \quad G_f] + [G_s \quad 0]T_{2sp}\}T_{1sp}\begin{bmatrix} x_1(k) \\ x_2(k) \end{bmatrix} \\ &= -\left\{[0 \quad G_f] + [G_s \quad 0]\begin{bmatrix} I & -\varepsilon P \\ 0 & I \end{bmatrix}\right\}\begin{bmatrix} I & 0 \\ L & I \end{bmatrix}\begin{bmatrix} x_1(k) \\ x_2(k) \end{bmatrix} \\ &= -[G_s \quad G_f - \varepsilon G_s P]\begin{bmatrix} I & 0 \\ L & I \end{bmatrix}\begin{bmatrix} x_1(k) \\ x_2(k) \end{bmatrix} \\ &= -\big[G_s + (G_f - \varepsilon G_s P)L \quad G_f - \varepsilon G_s P\big]\begin{bmatrix} x_1(k) \\ x_2(k) \end{bmatrix} \\ &= -G_{1eq}x_1(k) - G_{2eq}x_2(k) = -[G_{1eq} \quad G_{2eq}]x(k) = -G_{eq}x(k) \end{aligned} \tag{3.39}$$

with the equivalent feedback gains from the original state space variables equal to

$$G_{1eq} = G_s + \big(G_f - \varepsilon G_s P\big)L = G_s + G_{2eq}L, \quad G_{2eq} = G_f - \varepsilon G_s P \tag{3.40}$$

A unique solution of (3.28) exists for sufficiently small values of ε under Assumption 3.5. In that case, nonlinear algebraic equation (3.28) can be solved by performing fixed-point iterations on a system of linear algebraic equations as follows

$$(A_{22} - I)L^{(i+1)} = A_{21} + \varepsilon L^{(i)}\big(A_{11} - A_{12}L^{(i)}\big), \quad L^{(0)} = (A_{22} - I)^{-1}A_{21}, \quad i = 1, 2, \ldots \tag{3.41}$$

This algorithm converges with the rate of convergence of the order of ε, denoted by $O(\varepsilon)$, which means that after i iterations, the accuracy of $O(\varepsilon^i)$ is achieved, in other words, the accuracy is improved in each iteration by $O(\varepsilon)$. Another alternative for small values of ε is to use the Newton method for solving weakly nonlinear algebraic equation (3.28). It is well-known that the Newton method has a quadratic rate of convergence, assuming that the initial guess is very good, which is true in this case since $L^{(0)} = L + O(\varepsilon)$. Even, if ε is not sufficiently small, the eigenvector method can be used to solve (3.28). Having obtained L from (3.41), A_s, A_f, B_f can be calculated from (3.31), and G_f can be determined in Step 2 to set up the fast subsystem eigenvalues in the desired locations. Then, the algebraic equation (3.35) for P can be solved directly as a linear algebraic Sylvester equation (Chen 2012). A unique solution of the Sylvester algebraic equation exists if matrices $A_f - B_f G_f$ and $I + \varepsilon A_s$ have no eigenvalues in common (Chen 2012), that is, if $\lambda(A_f - B_f G_f) \neq 1 + \varepsilon\lambda(A_s)$, which can be easily satisfied by the appropriate selection of the closed-loop eigenvalues of the fast and slow subsystems. Moreover, the fast dynamics eigenvalues are far from the unit circle and close to the origin, so that in the case of systems with slow and fast modes, the algebraic Sylvester equation will always have a unique solution.

3.3.1 Example: A Steam Power System

In this example, we perform the eigenvalue assignment design in two stages and two-time scales by calculating independently slow and fast subsystem feedback gains using the reduced-order dynamic models and corresponding reduced-order matrices.

Consider a discrete-time model of a steam power system that can be studied as a system with slow and fast modes. The model state space matrices are as follows

$$\begin{bmatrix} x_1(k+1) \\ x_2(k+1) \\ x_3(k+1) \\ x_4(k+1) \\ x_5(k+1) \end{bmatrix} = \begin{bmatrix} 0.9150 & 0.0510 & 0.0380 & 0.0150 & 0.0380 \\ -0.0300 & 0.8890 & -0.0005 & 0.0460 & 0.1110 \\ -0.0060 & 0.4680 & 0.2470 & 0.0140 & 0.0480 \\ -0.7150 & -0.0220 & -0.0211 & 0.2400 & -0.0240 \\ -0.1480 & -0.0030 & -0.0040 & 0.0900 & 0.0260 \end{bmatrix} \begin{bmatrix} x_1(k) \\ x_2(k) \\ x_3(k) \\ x_4(k) \\ x_5(k) \end{bmatrix} + \begin{bmatrix} 0.0098 \\ 0.1220 \\ 0.0360 \\ 0.5620 \\ 0.1150 \end{bmatrix} u(k) = Ax(k) + Bu(k)$$

This system has two slow ($n_1 = 2$) and three fast ($n_2 = 3$) state variables. The small singular perturbation parameter is equal to $\varepsilon = 0.274$. Consistently to the singularly perturbed system notation used in (3.27), the corresponding matrices are given by

$$A_{11} = \begin{bmatrix} -0.3220 & 0.1932 \\ -0.1136 & -0.4205 \end{bmatrix}, \quad A_{12} = \begin{bmatrix} 0.1439 & 0.0568 & 0.1439 \\ -0.0019 & 0.1742 & 0.4205 \end{bmatrix}, \quad B_{11} = \begin{bmatrix} 0.0371 \\ 0.4621 \end{bmatrix}$$

$$A_{21} = \begin{bmatrix} -0.0060 & 0.4680 \\ -0.7150 & -0.0220 \\ -0.1480 & -0.0030 \end{bmatrix}, \quad A_{22} = \begin{bmatrix} 0.2470 & 0.0140 & 0.0480 \\ -0.0211 & 0.2400 & -0.0240 \\ -0.0040 & 0.0900 & 0.0260 \end{bmatrix}, \quad B_{22} = \begin{bmatrix} 0.0360 \\ 0.5620 \\ 0.1150 \end{bmatrix}$$

The solution of L-equation (3.28) is obtained by using the fixed-point algorithm (3.41), in which we run 23 iterations on the corresponding linear algebraic equations to obtain accuracy of $E^{(i)} = O(10^{-14})$, where the accuracy is defined as

$$E^{(i)} = L^{(i)}(I + \varepsilon A_{11}) - A_{22}L^{(i)} - \varepsilon L^{(i)}A_{12}L^{(i)} + A_{21} \approx 0$$

Note that this is consistent with the result

$$\varepsilon^{23} = (0.274)^{23} = 0.117 \times 10^{-14} = O\left(10^{-14}\right)$$

Having obtained $L \approx L^{(23)}$, we calculated A_s, A_f, B_f using (3.31), which produced

$$L = \begin{bmatrix} -0.0697 & -0.7171 \\ 1.0778 & -0.0746 \\ 0.2802 & -0.0269 \end{bmatrix} \quad \Rightarrow$$

$$A_s = \begin{bmatrix} -0.4135 & -0.3045 \\ -0.4194 & -0.3975 \end{bmatrix}, \quad A_f = \begin{bmatrix} 0.2447 & -0.0200 & -0.0342 \\ 0.0199 & 0.2527 & 0.0087 \\ 0.0067 & 0.0930 & 0.0337 \end{bmatrix}, \quad B_f = \begin{bmatrix} -0.0522 \\ 0.5633 \\ 0.1145 \end{bmatrix}$$

Assume that we want to place the closed-loop system slow eigenvalues at $\lambda_s^{desired} = \{-0.9,\ 0.8\}$ and to place the closed-loop fast eigenvalues at $\lambda_f^{desired} = \{-0.1 \pm j0.2,\ 0.1\}$. According to the presented two-stage feedback design, we first find the fast subsystem gain G_f (Step 2) that locates $\lambda\left(A_f - B_f G_f\right) = \lambda_{fast}^{desired}$. This leads (using the MATLAB function "place") to

$$G_f = \begin{bmatrix} -5.7207 & 0.2688 & 1.5832 \end{bmatrix}$$

In Step 3, we solve the Sylvester algebraic equation (3.35) directly using the MATLAB function "lyap," which produced

$$P = \begin{bmatrix} -0.3824 & -0.0580 & -0.0652 \\ -2.8153 & -0.0349 & 0.1905 \end{bmatrix}$$

In Step 4, we find the slow subsystem feedback gain using $1 + \varepsilon\lambda\left(A_s - \left(B_{11} - PB_f\right)G_s\right) = \lambda_{slow}^{desired}$ and the corresponding MATLAB function "place," which have produced

$$G_s = \begin{bmatrix} 27.8751 & 17.7130 \end{bmatrix}$$

These slow and fast gains G_s, G_f are obtained using calculations with the reduced-order matrices. The actual full-state feedback gain applied to the system is obtained in Step 5 as

$$\begin{aligned} G_{eq} &= \begin{bmatrix} G_{1eq} & G_{2eq} \end{bmatrix} = \begin{bmatrix} G_s + \left(G_f - \varepsilon G_s P\right)L & G_f - \varepsilon G_s P \end{bmatrix} \\ &= \begin{bmatrix} 28.4144 & 10.2606 & 10.2589 & 0.8587 & 1.1723 \end{bmatrix} \end{aligned}$$

It can be easily checked (using MATLAB) that indeed this gain produces the desired set of eigenvalues with the accuracy of $O(10^{-14})$.

If we use only the zero-order approximation for the corresponding feedback design (corresponding to $i = 0$), which was done in almost all papers dealing with the discrete-time singularly perturbed systems (Naidu and Calise 2001), we get a pretty poor approximation for the closed-loop eigenvalues given by

$$\lambda_{system}^{(0)} = \{-0.1676 \pm j0.1780, \quad 0.0128, \quad -0.6760, \quad 0.7984\}$$

Since in the proposed procedure the accuracy can be simply increased just by running additional iterations, we get easily

$$\lambda_{system}^{(1)} = \{-0.0971 \pm j0.2083, \quad 0.0925, \quad -0.8974, \quad 0.7991\}$$

and for $i = 6$, the accuracy of $O(10^{-4})$ is achieved, that is

$$\lambda_{system}^{(6)} = \{-0.100025 \pm j0.199981, \quad 0.100002, \quad -0.899967, \quad 0.800002\}$$

As indicated before, in this example, $i = 23$ iterations are needed for the accuracy of $O(10^{-14})$.

It should be emphasized that the equation (3.35) for P is solved exactly using MATLAB function "lyap." If that equation was solved approximately (like the L-equation (3.28) using the corresponding fixed-point iterations similar to (3.41)), the results obtained for the approximate eigenvalues would be worse. For example, using

$$P^{(0)} = -\left(A_{12} - B_{11}G_f\right)\left[I - \left(A_f - B_fG_f\right)\right]^{-1} = P + O(\varepsilon) \tag{3.42}$$

and $L^{(0)}$ in the approximate procedure to locate the eigenvalues would produce

$$\lambda_{system}^{(0,0)} = \{-0.1183 \pm j0.1771, \quad 0.1272, \quad 0.8110, \quad -1.0433\}$$

which is not only far away from the desired eigenvalues

$$\lambda_f^{desired} = \{-0.1 \pm j0.2, \ 0.1\} \quad \text{and} \quad \lambda_s^{desired} = \{-0.9, \ 0.8\}$$

but also produces an unstable discrete-time closed-loop system. Hence, the $O(\varepsilon)$ accuracy might lead to inaccurate and/or unacceptable results. Fortunately, the two-stage feedback design provides a methodology to achieve a very accurate placement of the desired eigenvalues using only operations with reduced-order submatrices corresponding to the slow and fast subsystems.

It is possible that for some systems the MATLAB function "lyap" (used for solving the Sylvester algebraic equation) produces only a limited accuracy for the Sylvester algebraic equation. In such cases, solving the Sylvester algebraic equation of singularly perturbed systems, (3.35), can be done iteratively, like the L-equation given in (3.41), which can produce very high accuracy (assuming that the singular perturbation parameter ε is sufficiently small).

In Figs. 3.1, 3.2, 3.3, and 3.4, we have plotted the slow and fast subsystem step and impulse responses of the considered steam power system under the feedback

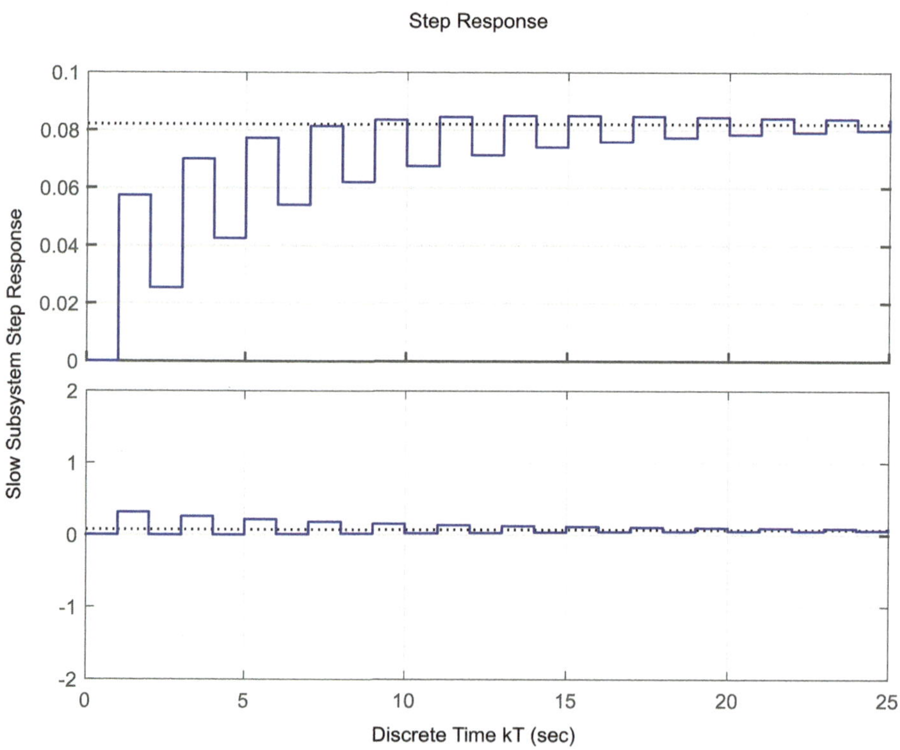

Fig. 3.1 Slow subsystem of the steam power system step response

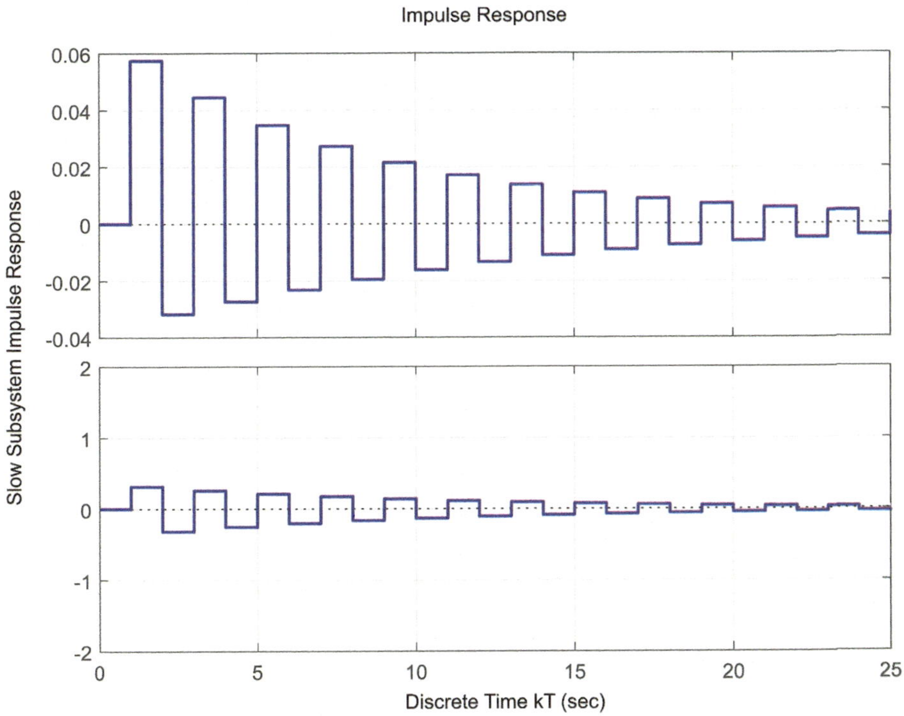

Fig. 3.2 Slow subsystem of the steam power system impulse response

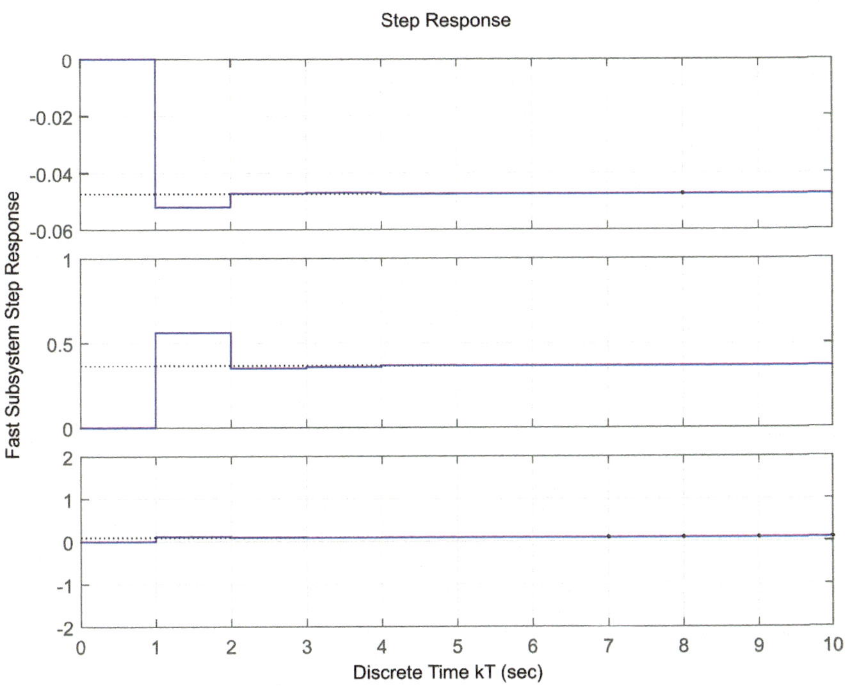

Fig. 3.3 Fast subsystem of the steam power system step response

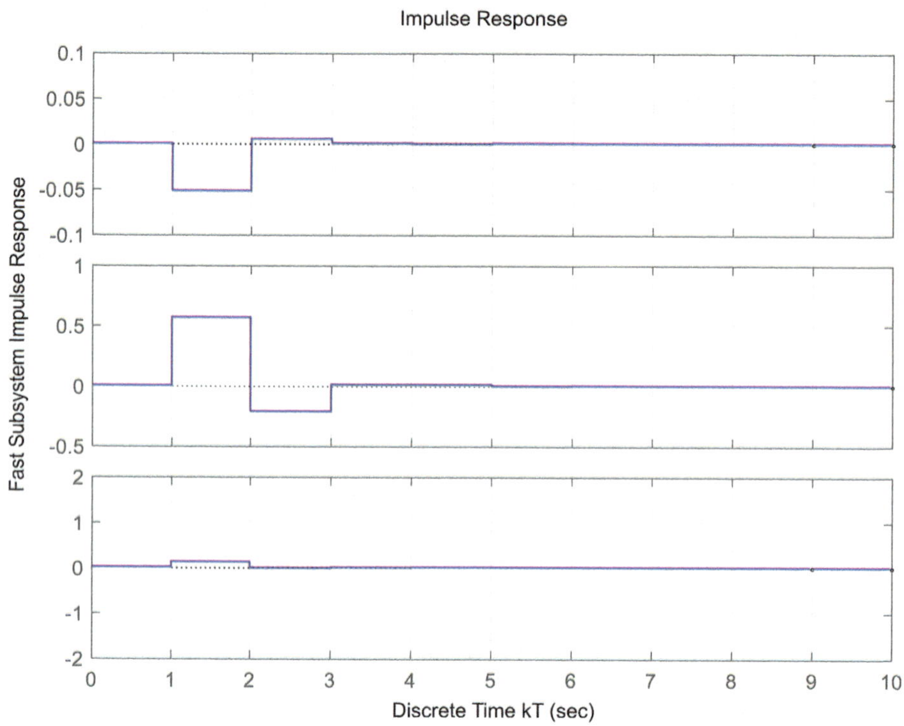

Fig. 3.4 Fast subsystem of the steam power system impulse response

control law that assigned the desired slow and fast closed-loop eigenvalues. It can be seen from these figures that the fast subsystem step and impulse responses go to their steady-state values in three to four discrete-time steps, whereas the corresponding responses of the slow subsystem need much more time (more than 25 discrete-time steps) to reach their steady-state values.

3.4 Notes

In this chapter we have fully derived all design equations of two-stage feedback controller algorithms for both formulations of discrete-time singularly perturbed time-invariant linear dynamic systems. In Chap. 5, we will present an extension of the two-stage to the three-stage feedback controller design of general discrete-time time-invariant linear dynamic systems. However, an extension of the three-stage design to the *three-time scale singularly perturbed discrete-time systems* at the present time is not possible since there are at least four formulations of three-time scale singularly perturbed systems, and it is not clear at the present time which one is the best suited for the considered three-stage feedback controller design. Hence, that problem remains as an interesting and challenging area for future research. More about it and its challenges and opportunities will be said in the second part of Chap. 5.

Presentation in Sects. 3.1–3.3 is based on the papers Radisavljevic-Gajic (2015a, b). Permissions for the use of such material in this research monograph were granted to us by the American Society of Mechanical Engineers (ASME) for two journal papers by Radisavljevic-Gajic (2015a, b), published in *Transactions of ASME Journal of Dynamic Systems Measurement and Control.*

Chapter 4
Three-Stage Continuous-Time Feedback Controller Design

In this chapter, the results of two-stage continuous-time feedback controller design from Chap. 2 are extended to the three-stage feedback controller design. This facilitates control of three subsets of system state variables representing three artificial or natural subsystems of a system under consideration. The presentation follows the recent papers of Radisavljevic-Gajic and Milanovic (2016) and Radisavljevic-Gajic et al. (2017). The new technique introduces simplicity and requires only solutions of reduced-order subsystem level algebraic equations for the design of appropriate local controllers. The local feedback controllers are combined to form a global controller for the system under consideration. The technique presented facilitates designs of independent full-state feedback controllers at the subsystem levels. Different types of local controllers, for example, eigenvalue assignment, robust, optimal in some sense (L_1, H_2, H_∞,...), observer-driven, Kalman filter-driven controllers, may be used to control different subsystems. This feature has not been available for any other known linear feedback controller design technique.

In the second part of the chapter, we specialize the results obtained to the three-time scale linear control systems (singularly perturbed control systems) that have natural decomposition into slow, fast, and very fast subsystems. The proposed technique eliminates numerical ill-conditioning of the original three-time scale singularly perturbed linear systems. The newly proposed three-stage feedback controller design is demonstrated on the eighth-order proton exchange membrane fuel cell model.

V. Radisavljević-Gajić et al., *Multi-Stage and Multi-Time Scale Feedback Control of Linear Systems with Applications to Fuel Cells*, Mechanical Engineering Series,
https://doi.org/10.1007/978-3-030-10389-7_4

4.1 Introduction

The technique to be presented requires that the system subsystems are clearly identified. The system partitioning into several subsystems can be done using several methods: (1) based on the physical nature of subsystems (system natural decomposition); (2) according to the conditions that must be satisfied such that the partitioned system is feasible for the three-stage feedback design; (3) based on mathematical conditions that must be satisfied to solve the corresponding design equations; (4) control needs (which parts of the system should be independently controlled via best suited local controllers); and/or (5) grouping of the state space variables such that subsystems satisfy control-oriented assumptions, for example, needed for the design of local optimal controllers and filters such as controllability (stabilizability) and/or observability (detectability) conditions. It is possible that the system stabilizability-detectability conditions are not satisfied so that the linear system linear-quadratic optimal controller cannot be designed, but the stabilizability-detectability conditions can be satisfied for local subsystems and locally optimal linear-quadratic controllers can be designed.

The power of the three-stage feedback design is in the following:

(a) Different types of controllers (optimal, eigenvalue assignment, robust, reliable, etc.) can be designed for different parts of the system (subsystems) using corresponding feedback gains obtained by performing calculations (design) only with subsystem (reduced-order) matrices.
(b) Local subsystem feedback gains that control local subsystems are compounded into one full-state feedback gain via a single formula, leading to a unified feedback controller for the system under consideration.
(c) Computational requirements are drastically reduced (especially for two- and three-time scale systems) since computations are done with matrices of reduced dimensions corresponding to subsystems.
(d) Very high accuracy can be achieved since numerical ill-conditioning of higher-order matrices can be eliminated and computations performed with well-conditioned lower-order matrices.
(e) The design can be extended for the development of corresponding three-stage observers (Sinha 2007; Chen 2012) and filters, as well as for observer- and filter-driven controllers (it can be also extended to stochastic systems) including their three-time scale counterparts.
(f) The design is independent for each local subsystem so that it provides flexibility for the development of partial full-state feedback and partial output feedback controllers including linear-quadratic optimal controllers.
(g) Robustness and reliability can be easily facilitated by using three- and multistage designs, and the feedback control loop security can be improved, which appears to be a very important feature these days, especially for cyber physical systems and computer/communication networks.

As an important application of the proposed methodology, we specialize it to large-scale linear control systems with slow and fast modes (multi-time scale feedback systems, singularly perturbed control systems) (Kokotovic et al. 1999; Naidu and Calise 2001; Gajic and Lim 2001; Dimitriev and Kurina 2006; Kuehn 2015), for which the proposed types of designs seem to be very well suited. Many systems in mechanical and aerospace engineering possess the singularly perturbed structure (Hsiao et al. 2001; Naidu and Calise 2001; Chen et al. 2002; Shapira and Ben-Asher 2004; Demetriou and Kazantzis 2005; Wang and Ghorbel 2006; Amjadifard et al. 2011; Kuehn 2015) due to the presence of small and large time constants, small masses, small moments of inertia, and small stiffness coefficients, which causes clustering of the system eigenvalues into two or several disjoint groups close to the imaginary axis (slow eigenvalues) and farther from the imaginary axis (fast eigenvalues) and very far from the imaginary axis (very fast eigenvalues). Other locations of linear system eigenvalues in the complex plane that produce three-time scale dynamics are possible.

4.2 Three-Stage Design of Continuous-Time Feedback Controllers

The efficient methods for the two-stage linear feedback controller designs have been presented in Chaps. 2 and 3, respectively, in continuous- and discrete-time domains. In this section, we will extend the continuous-time results of Chap. 2 to the three-stage feedback control design and in the follow-up section show how the newly obtained results can be efficiently used for feedback control of three-time scale linear control systems (singularly perturbed linear control systems).

Consider a continuous-time, time-invariant, linear system represented in its partitioned form by

$$\frac{dx(t)}{dt} = \begin{bmatrix} \dfrac{dx_{\mathrm{I}}(t)}{dt} \\ \dfrac{dx_{\mathrm{II}}(t)}{dt} \\ \dfrac{dx_{\mathrm{III}}(t)}{dt} \end{bmatrix} = \begin{bmatrix} A_{11} & A_{12} & A_{13} \\ A_{21} & A_{22} & A_{23} \\ A_{31} & A_{32} & A_{33} \end{bmatrix} \begin{bmatrix} x_{\mathrm{I}}(t) \\ x_{\mathrm{II}}(t) \\ x_{\mathrm{III}}(t) \end{bmatrix} + \begin{bmatrix} B_{11} \\ B_{22} \\ B_{33} \end{bmatrix} u(t) = Ax(t) + Bu(t)$$

$$y(t) = \begin{bmatrix} C_{11} & C_{22} & C_{33} \end{bmatrix} \begin{bmatrix} x_{\mathrm{I}}(t) \\ x_{\mathrm{II}}(t) \\ x_{\mathrm{III}}(t) \end{bmatrix} \tag{4.1}$$

where $x(t) \in \mathrm{R}^n$, $x_{\mathrm{I}}(t) \in \mathrm{R}^{n_1}$, $x_{\mathrm{II}}(t) \in \mathrm{R}^{n_2}$, and $x_{\mathrm{III}}(t) \in \mathrm{R}^{n_3}$, $n = n_1 + n_2 + n_3$, are state variables, $u(t) \in \mathrm{R}^m$ is the control input vector, $y(t) \in \mathrm{R}^p$ is the vector of system measurements, and A_{ij}, B_{ii}, and C_{ii}, $i,j = 1,2,3$, are constant matrices of appropriate dimensions. Matrices A_{11}, A_{22}, and A_{33} define subsystems of dimensions n_1, n_2, and n_3, respectively, corresponding to the state variables $x_{\rm I}(t)$, $x_{\rm II}(t)$, and $x_{\rm III}(t)$. Matrices A_{ij}, $i,j = 1,2,3$, $i \neq j$ define couplings between the subsystems.

The steps of the three-stage feedback controller design are presented in the follow-up of this section. In the introductory stage, we map the system into the appropriate coordinates via several transformations in order to achieve the upper block triangular form that facilitates the three-stage feedback controller design. The presentation is done mostly by following the work of Radisavljevic-Gajic et al. (2017).

We start with the following change of variables applied to the original system defined in (4.1)

$$\eta_3(t) = L_1 x_{\rm I}(t) + L_2 x_{\rm II}(t) + x_{\rm III}(t) \tag{4.2}$$

which produces the dynamic equation for $\eta_3(t)$ as

$$\begin{aligned}\frac{d\eta_3(t)}{dt} &= (A_{33} + L_1 A_{13} + L_2 A_{23})\eta_3(t) + (B_{33} + L_1 B_{11} + L_2 B_{22})u(t) \\ &\quad + f_{31}(L_1, L_2)x_{\rm I}(t) + f_{32}(L_1, L_2)x_{\rm II}(t) = A_3\eta_3(t) + B_3 u(t)\end{aligned} \tag{4.3}$$

$$A_3 = A_{33} + L_1 A_{13} + L_2 A_{23}, \quad B_3 = B_{33} + L_1 B_{11} + L_2 B_{22}$$

Elimination of coupling terms in (4.3) is achieved by assuming that the following systems of algebraic equations have real solutions:

$$\begin{aligned} f_{31}(L_1, L_2) &= L_1 A_{11} + L_2 A_{21} + A_{31} - (L_1 A_{13} + L_2 A_{23} + A_{33})L_1 = 0 \\ f_{32}(L_1, L_2) &= L_1 A_{12} + L_2 A_{22} + A_{32} - (L_1 A_{13} + L_2 A_{23} + A_{33})L_2 = 0 \end{aligned} \tag{4.4}$$

Solving coupled nonlinear algebraic equations (4.4), in general, is not an easy task. However, we will show in the next section that for the three-time scale linear systems, the corresponding algebraic equations have much simpler forms so that can be easily solved numerically using the fixed-point iterations as systems of *linear algebraic equations*.

Eliminating $x_{\rm III}(t)$ from (4.1), the differential equations for $x_{\rm I}(t)$ and $x_{\rm II}(t)$ become

$$\frac{dx_{\rm I}(t)}{dt} = (A_{11} - A_{13}L_1)x_{\rm I}(t) + (A_{12} - A_{13}L_2)x_{\rm II}(t) + A_{13}\eta_3(t) + B_{11}u(t) \tag{4.5}$$

and

$$\frac{dx_{\rm II}(t)}{dt} = (A_{21} - A_{23}L_1)x_{\rm I}(t) + (A_{22} - A_{23}L_2)x_{\rm II}(t) + A_{23}\eta_3(t) + B_{22}u(t) \tag{4.6}$$

Since the $\eta_3(t)$-subsystem (4.3) is isolated, one can design a feedback controller for it using $u(t) = -G_3\eta_3(t)+v(t)$, with $v(t)$ representing an input signal that can be used to control the first two subsystems. This strategy was used in the two-stage feedback controller design in Chap. 2. However, *in this chapter due to a more complex three-stage design, we will first achieve an upper block triangular structure for the overall system and then show how to design independently local feedback controllers*.

Now, we introduce the second change of variables to remove dynamics of the first subsystem from the second subsystem

$$\eta_2(t) = L_3 x_{\mathrm{I}}(t) + x_{\mathrm{II}}(t) \tag{4.7}$$

This change of variables leads to

$$\begin{aligned}\frac{d\eta_2(t)}{dt} &= [A_{22} - A_{23}L_2 + L_3(A_{12} - A_{13}L_2)]\eta_2(t) \\ &\quad + (A_{23} + L_3A_{13})\eta_3(t) + f_{21}(L_1, L_2, L_3)x_{\mathrm{I}}(t) + (B_{22} + L_3B_{11})u(t) \\ &= A_2\eta_2(t) + (A_{23} + L_3A_{13})\eta_3(t) + B_2u(t)\end{aligned} \tag{4.8}$$

$$A_2 = A_{22} - A_{23}L_2 + L_3(A_{12} - A_{13}L_2), \quad B_2 = B_{22} + L_3B_{11}$$

where L_3 satisfies the following algebraic equation

$$L_3(A_{11} - A_{13}L_1) - (A_{22} - A_{23}L_2)L_3 - L_3(A_{12} - A_{13}L_2)L_3 + (A_{21} - A_{23}L_1) = 0 \tag{4.9}$$

Assuming that L_1 and L_2 are previously obtained from (4.4), the matrix L_3 in fact satisfies the nonsymmetric, nonsquare algebraic Riccati equation that was previously studied in Medanic (1982) and Gao and Bai (2010).

After transformation (4.7), the first subsystem in the new coordinates becomes

$$\begin{aligned}\frac{dx_{\mathrm{I}}(t)}{dt} &= [(A_{11} - A_{13}L_1) - (A_{12} - A_{13}L_2)L_3]x_{\mathrm{I}}(t) \\ &\quad + (A_{12} - A_{13}L_2)\eta_2(t) + A_{13}\eta_3(t) + B_{11}u(t) \\ &= A_1x_{\mathrm{I}}(t) + (A_{12} - A_{13}L_2)\eta_2(t) + A_{13}\eta_3(t) + B_{11}u(t)\end{aligned} \tag{4.10a}$$

where

$$A_1 = (A_{11} - A_{13}L_1) - (A_{12} - A_{13}L_2)L_3 \tag{4.10b}$$

Transformations (4.2) and (4.7) relate the original and new state variables as follows

$$
\begin{aligned}
\begin{bmatrix} x_{\mathrm{I}}(t) \\ \eta_2(t) \\ \eta_3(t) \end{bmatrix} &= T_1 \begin{bmatrix} x_{\mathrm{I}}(t) \\ x_{\mathrm{II}}(t) \\ x_{\mathrm{III}}(t) \end{bmatrix} = \begin{bmatrix} I & 0 & 0 \\ L_3 & I & 0 \\ L_1 & L_2 & I \end{bmatrix} \begin{bmatrix} x_{\mathrm{I}}(t) \\ x_{\mathrm{II}}(t) \\ x_{\mathrm{III}}(t) \end{bmatrix} \\
\begin{bmatrix} x_{\mathrm{I}}(t) \\ x_{\mathrm{II}}(t) \\ x_{\mathrm{III}}(t) \end{bmatrix} &= T_1^{-1} \begin{bmatrix} x_{\mathrm{I}}(t) \\ \eta_2(t) \\ \eta_3(t) \end{bmatrix} = \begin{bmatrix} I & 0 & 0 \\ -L_3 & I & 0 \\ -L_1 + L_2 L_3 & -L_2 & I \end{bmatrix} \begin{bmatrix} x_{\mathrm{I}}(t) \\ \eta_2(t) \\ \eta_3(t) \end{bmatrix}
\end{aligned}
\tag{4.11}
$$

This similarity transformation maps the original system (4.1) into the upper block triangular form represented by equations (4.3), (4.8), and (4.10), whose state space form is given by

$$
\begin{aligned}
\begin{bmatrix} \dfrac{dx_{\mathrm{I}}(t)}{dt} \\ \dfrac{d\eta_2(t)}{dt} \\ \dfrac{d\eta_3(t)}{dt} \end{bmatrix} &= \begin{bmatrix} A_1 & A_{12} - A_{13}L_2 & A_{13} \\ 0 & A_2 & A_{23} + L_3 A_{13} \\ 0 & 0 & A_3 \end{bmatrix} \begin{bmatrix} x_{\mathrm{I}}(t) \\ \eta_2(t) \\ \eta_3(t) \end{bmatrix} + \begin{bmatrix} B_{11} \\ B_2 \\ B_3 \end{bmatrix} u(t) \\
&= \bar{A} \begin{bmatrix} x_{\mathrm{I}}(t) \\ \eta_2(t) \\ \eta_3(t) \end{bmatrix} + \bar{B} u(t) \\
A_1 &= (A_{11} - A_{13}L_1) - (A_{12} - A_{13}L_2)L_3 \\
A_2 &= A_{22} - A_{23}L_2 + L_3(A_{12} - A_{13}L_2), \quad B_2 = B_{22} + L_3 B_{11} \\
A_3 &= A_{33} + L_1 A_{13} + L_2 A_{23}, \quad B_3 = B_{33} + L_1 B_{11} + L_2 B_{22}
\end{aligned}
\tag{4.12}
$$

$$
\begin{aligned}
y(t) &= (C_{11} - C_{22}L_3 - C_{33}L_1)x_{\mathrm{I}}(t) + (C_{22} - C_{33}L_2)\eta_2(t) + C_{33}\eta_3(t) \\
&= C_1 x_{\mathrm{I}}(t) + C_2 \eta_2(t) + C_{33}\eta_3(t) = \bar{C} \begin{bmatrix} x_{\mathrm{I}}(t) \\ \eta_2(t) \\ \eta_3(t) \end{bmatrix} \\
C_1 &= C_{11} - C_{22}L_3 - C_{33}L_1 + C_{33}L_2L_3, \quad C_2 = C_{22} - C_{33}L_2
\end{aligned}
$$

Having transformed the original system into an upper block triangular form, we can start the design procedure of feedback controllers. We will indicate intermediate steps needed to make these designs independent of each other. If one plans to design *optimal* linear-quadratic feedback controllers, the transformed system needs to satisfy the controllability and observability conditions or its weaker variants stabilizability and detectability conditions. In the next remark, we clarify this issue by imposing these conditions on the original system (4.1) and show that they hold for the transformed system (4.12) as well.

Remark 4.1: Stabilizability-Detectability Invariance Under Similarity Transformation If the original system (4.1) is controllable and observable, then the system (4.12) is also controllable and observable since the original and new coordinates are related via the similarity transformation (4.11), which is the very well-known result; see, for example, Chen (2012). The same invariance result can be established for the stabilizability and detectability conditions, namely, if the original system (4.1) is stabilizable and detectable, then the system (4.12) is also stabilizable and detectable. Since the *stabilizability and detectability invariance* result hardly can be found in the literature, we provide here its simple proof using the Popov-Belevitch-Hautus eigenvector stabilizability test (Zhou and Doyle 1998). In the following, we prove that stabilizability is invariant under a similarity transformation.

Note that using a similarity transformation, we have in the new coordinates the following relationships $\bar{A} = T^{-1}AT$ and $\bar{B} = T^{-1}B$. By the Popov-Belevitch-Hautus test, the pair (A, B) is stabilizable if for any *unstable eigenvalue* λ_i and the corresponding left eigenvector of A, that is, $w_i^* A = \lambda_i w^i i^*$, the following holds $w_i^* B \neq 0$, indicating that there is no a left eigenvector of A orthogonal to the column of matrix B. Since $A = T\bar{A}T^{-1}$ and $B = T\bar{B}$, then $w_i^* T\bar{A}T^{-1} = \lambda_i w^i i^*$ or $(Tw_i)^* \bar{A} = \lambda_i (Tw_i)^i *$ indicates that if w_i^* is a left eigenvector of A, then $(Tw_i)^*$ is a left eigenvector of $\bar{A}$. Using this fact, we have that $w_i^* B \neq 0$ implies $w_i^* TT^{-1}B \neq 0$ or $(Tw_i)^* \bar{B} \neq 0$, which proves that the pair $(\bar{A}, \bar{B})$ is also stabilizable.

Similarly, using the Popov-Belevitch-Hautus eigenvector detectability test, we can establish that *detectability of the pair* (A, C) *is equivalent to detectability of the pair* $(\bar{A}, \bar{C})$.

Having obtained the upper block triangular form in (4.12), now we start the three-stage feedback controller design procedure by following the work of Radisavljevic-Gajic et al. (2017), in which we isolate every subsystem and identify the control input that independently of other subsystems solely controls the isolated subsystem.

Stage 1
Apply feedback control

$$u(t) = -G_3\eta_3(t) + v(t)$$

to the η-subsystem, which leads to

$$\begin{bmatrix} \dfrac{dx_1(t)}{dt} \\ \dfrac{d\eta_2(t)}{dt} \\ \dfrac{d\eta_3(t)}{dt} \end{bmatrix} = \begin{bmatrix} A_1 & A_{12} - A_{13}L_2 & A_{13} - B_{11}G_3 \\ 0 & A_2 & A_{23} + L_3A_{13} - B_2G_3 \\ 0 & 0 & A_3 - B_3G_3 \end{bmatrix} \begin{bmatrix} x_1(t) \\ \eta_2(t) \\ \eta_3(t) \end{bmatrix} + \begin{bmatrix} B_{11} \\ B_2 \\ B_3 \end{bmatrix} v(t) \tag{4.13}$$

Stage 2

To isolate the second continuous-time linear dynamic subsystem, we apply the change of the state space variables as

$$\xi_2(t) = \eta_2(t) - P_3\eta_3(t) \tag{4.14}$$

which produces

$$\begin{bmatrix} \frac{dx_I(t)}{dt} \\ \frac{d\xi_2(t)}{dt} \\ \frac{d\eta_3(t)}{dt} \end{bmatrix} = \begin{bmatrix} A_1 & A_{12} - A_{13}L_2 & A_{13} - B_{11}G_3 + (A_{12} - A_{13}L_2)P_3 \\ 0 & A_2 & 0 \\ 0 & 0 & A_3 - B_3G_3 \end{bmatrix} \begin{bmatrix} x_I(t) \\ \xi_2(t) \\ \eta_3(t) \end{bmatrix} + \begin{bmatrix} B_{11} \\ B_2 - P_3B_3 \\ B_3 \end{bmatrix} v(t) \tag{4.15}$$

where P_3 satisfies the following algebraic equation

$$A_2P_3 - P_3(A_3 - B_3G_3) + A_{23} + L_3A_{13} - B_2G_3 = 0 \tag{4.16}$$

Note that (4.16) is the Sylvester algebraic equation, and its unique solution exists under the following assumption (Chen 2012).

Assumption 4.1 Matrices A_2 and $A_3 - B_3G_3$ have no eigenvalues in common.

This assumption is easily satisfied since $A_3 - B_3G_3$ is the feedback matrix of the third subsystem, and the linear algebraic equation (4.16) can be solved.

In Stage 2, we apply feedback control to the second subsystem as

$$v(t) = -G_2\xi_2(t) + w(t)$$

which leads to

$$\begin{bmatrix} \frac{dx_I(t)}{dt} \\ \frac{d\xi_2(t)}{dt} \\ \frac{d\eta_3(t)}{dt} \end{bmatrix} = \begin{bmatrix} A_1 & A_{12} - A_{13}L_2 - B_{11}G_2 & A_{13} - B_{11}G_3 + (A_{12} - A_{13}L_2)P_3 \\ 0 & A_2 - (B_2 - P_3B_3)G_2 & 0 \\ 0 & -B_3G_2 & A_3 - B_3G_3 \end{bmatrix} \begin{bmatrix} x_I(t) \\ \xi_2(t) \\ \eta_3(t) \end{bmatrix} + \begin{bmatrix} B_{11} \\ B_2 - P_3B_3 \\ B_3 \end{bmatrix} w(t) \tag{4.17}$$

Stage 3

Now we need to isolate the first subsystem and remove its coupling from the second and third subsystems. Before we proceed with this task, we first introduce the simplifying notation and present differential equation (4.17) in the form

$$\begin{bmatrix} \dfrac{dx_{\mathrm{I}}(t)}{dt} \\ \dfrac{d\xi_2(t)}{dt} \\ \dfrac{d\eta_3(t)}{dt} \end{bmatrix} = \begin{bmatrix} \alpha_{11} & \alpha_{12} & \alpha_{13} \\ 0 & \alpha_{22} & 0 \\ 0 & \alpha_{32} & \alpha_{33} \end{bmatrix} \begin{bmatrix} x_{\mathrm{I}}(t) \\ \xi_2(t) \\ \eta_3(t) \end{bmatrix} + \begin{bmatrix} \beta_1 \\ \beta_2 \\ \beta_3 \end{bmatrix} w(t) \tag{4.18a}$$

where

$$\begin{aligned} &\alpha_{11} = A_1, \quad \alpha_{12} = A_{12} - A_{13}L_2 - B_{11}G_2, \\ &\alpha_{13} = A_{13} - B_{11}G_3 + (A_{12} - A_{13}L_2)P_3, \quad \alpha_{22} = A_2 - (B_2 - P_3B_3)G_2 \\ &\alpha_{32} = -B_3G_2, \quad \alpha_{33} = A_3 - B_3G_3 \end{aligned} \tag{4.18b}$$

To isolate the first subsystem, we apply another change of state variables as follows

$$\xi_1(t) = x_{\mathrm{I}}(t) - P_1\xi_2(t) - P_2\eta_3(t) \tag{4.19}$$

This change of variables modifies (4.18) into

$$\begin{bmatrix} \dfrac{d\xi_1(t)}{dt} \\ \dfrac{d\xi_2(t)}{dt} \\ \dfrac{d\eta_3(t)}{dt} \end{bmatrix} = \begin{bmatrix} \alpha_{11} & 0 & 0 \\ 0 & \alpha_{22} & 0 \\ 0 & \alpha_{32} & \alpha_{33} \end{bmatrix} \begin{bmatrix} \xi_1(t) \\ \xi_2(t) \\ \eta_3(t) \end{bmatrix} + \begin{bmatrix} \beta_1 - P_1\beta_2 - P_2\beta_3 \\ \beta_2 \\ \beta_3 \end{bmatrix} w(t) \tag{4.20}$$

where P_1 and P_2 satisfy the following system of two linear matrix algebraic equations

$$\alpha_{11}P_1 - P_1\alpha_{22} - P_2\alpha_{32} + \alpha_{12} = 0 \tag{4.21}$$

$$\alpha_{11}P_2 - P_2\alpha_{33} + \alpha_{13} = 0 \tag{4.22}$$

The solution for P_2 can be directly obtained from (4.22) as a solution of the algebraic Sylvester equation. Having obtained the solution for P_2, equation (4.21) is another Sylvester equation that directly produces P_1. The unique solutions of these Sylvester equations exist under the following assumptions.

Assumption 4.2a (Needed for (4.22)) Matrices α_{11} and α_{33} have no eigenvalues in common.

Assumption 4.2b (Needed for (4.21)) Matrices α_{11} and α_{22} have no eigenvalues in common.

Since α_{22} and α_{33} are feedback matrices for the second and the third subsystems, both assumptions can be easily satisfied by choosing appropriate feedback gains.

Now the local feedback control input can be applied to the first subsystem as $w(t) = -G_1\xi_1(t)$

$$\begin{bmatrix} \frac{d\xi_1(t)}{dt} \\ \frac{d\xi_2(t)}{dt} \\ \frac{d\eta_3(t)}{dt} \end{bmatrix} = \begin{bmatrix} \alpha_{11} - (\beta_1 - P_1\beta_2 - P_2\beta_3)G_1 & 0 & 0 \\ -\beta_2 G_1 & \alpha_{22} & 0 \\ -\beta_3 G_1 & \alpha_{32} & \alpha_{33} \end{bmatrix} \begin{bmatrix} \xi_1(t) \\ \xi_2(t) \\ \eta_3(t) \end{bmatrix} \tag{4.23}$$

The obtained system matrix is lower block triangular so its closed-loop eigenvalues are the union of the closed-loop eigenvalues of the individual subsystems.

We can now relate the new and original state variables via a similarity transformation. The state coordinates defined in (4.12) and (4.23) are related via state transformations (4.14) and (4.19). These relationships can be put in a compact matrix form providing a unique transformation to the original state variables as

$$\begin{aligned} \begin{bmatrix} \xi_1(t) \\ \xi_2(t) \\ \eta_3(t) \end{bmatrix} &= \begin{bmatrix} I & -P_1 & -P_2 \\ 0 & I & 0 \\ 0 & 0 & I \end{bmatrix} \begin{bmatrix} x_{\mathrm{I}}(t) \\ \xi_2(t) \\ \eta_3(t) \end{bmatrix} = T_3 \begin{bmatrix} x_{\mathrm{I}}(t) \\ \xi_2(t) \\ \eta_3(t) \end{bmatrix} \\ &= T_3 \begin{bmatrix} I & 0 & 0 \\ 0 & I & -P_3 \\ 0 & 0 & I \end{bmatrix} \begin{bmatrix} x_{\mathrm{I}}(t) \\ \eta_2(t) \\ \eta_3(t) \end{bmatrix} = T_3T_2 \begin{bmatrix} x_{\mathrm{I}}(t) \\ \eta_2(t) \\ \eta_3(t) \end{bmatrix} \\ &= T_3T_2 \begin{bmatrix} I & 0 & 0 \\ L_3 & I & 0 \\ L_1 & L_2 & I \end{bmatrix} \begin{bmatrix} x_{\mathrm{I}}(t) \\ x_{\mathrm{II}}(t) \\ x_{\mathrm{III}}(t) \end{bmatrix} = T_3T_2T_1 \begin{bmatrix} x_{\mathrm{I}}(t) \\ x_{\mathrm{II}}(t) \\ x_{\mathrm{III}}(t) \end{bmatrix} = T \begin{bmatrix} x_{\mathrm{I}}(t) \\ x_{\mathrm{II}}(t) \\ x_{\mathrm{III}}(t) \end{bmatrix} \\ &= \begin{bmatrix} I - P_1L_3 - P_2L_1 + P_1P_3L_1 & -P_1 - P_2L_2 + P_1P_3L_2 & P_1P_3 - P_2 \\ L_3 - P_3L_1 & I - P_3L_2 & -P_3 \\ L_1 & L_2 & I \end{bmatrix} \begin{bmatrix} x_{\mathrm{I}}(t) \\ x_{\mathrm{II}}(t) \\ x_{\mathrm{III}}(t) \end{bmatrix} \end{aligned} \tag{4.24}$$

The feedback control signal applied to the system in the transformed coordinates is given by

$$\begin{aligned} u(\xi_1(t), \xi_2(t), \eta_3(t)) &= -G_1\xi_1(t) - G_2\xi_2(t) - G_3\eta_3(t) \\ &= -\begin{bmatrix} G_1 & G_2 & G_3 \end{bmatrix} \begin{bmatrix} \xi_1(t) \\ \xi_2(t) \\ \eta_3(t) \end{bmatrix} = -\begin{bmatrix} G_1 & G_2 & G_3 \end{bmatrix} T \begin{bmatrix} x_{\mathrm{I}}(t) \\ x_{\mathrm{II}}(t) \\ x_{\mathrm{III}}(t) \end{bmatrix} \end{aligned} \tag{4.25}$$

Using the state transformation (4.24), we can obtain the feedback control signal in the original coordinates

$$u(x_{\mathrm{I}}(t), x_{\mathrm{II}}(t), x_{\mathrm{III}}(t)) = -G_{1eq}x_{\mathrm{I}}(t) - G_{2eq}x_{\mathrm{II}}(t) - G_{3eq}x_{\mathrm{III}}(t) \tag{4.26}$$

with the equivalent feedback gains given by

$$\begin{aligned} G_{1eq} &= G_1(I - P_1L_3 - P_2L_1 + P_1P_3L_1) + G_2(L_3 - P_3L_1) + G_3L_1 \\ G_{2eq} &= G_1(P_1P_3L_2 - P_1 - P_2L_2) + G_2(I - P_3L_2) + G_3L_2 \\ G_{3eq} &= G_1(P_1P_3 - P_2) - G_2P_3 + G_3 \end{aligned} \tag{4.27}$$

The three-stage controller design gets simplified in the case of three-time scale systems that have a natural decomposition into slow, fast, and very fast subsystems. Namely, the corresponding design equations get simpler forms, and due to the presence of two small parameters (that identify the time scales), the L- and P-equations can be easily solved.

4.3 Three-Stage Three-Time Scale Linear Control Systems

Linear multi-time scale control systems are particularly well suited for the considered three-stage feedback design. For this class of systems, in general, numerical ill-conditioning appears if one attempts to design a linear feedback controller directly using the entire (full-order) system. As indicated in the introductory section, singularly perturbed control systems have numerous applications in all areas of engineering and sciences. In this section, further design simplifications will be achieved by specializing the proposed design from Section 4.2 to singularly perturbed linear systems so that only solutions of *linear algebraic equations* will be required. Moreover, the digital implementation of the corresponding controllers will allow different sampling periods to be used for the slow (large sampling period), fast (small sampling period), and very fast (very small sampling period) controllers. Otherwise, without the three-stage design, the whole system digital controller will require a very small sampling period (very large sampling rate).

Several multi-time scales are present in many real physical systems that have components of different nature. For example, advanced heavy water reactor (Smimjith et al. 2011a, b; Munje et al. 2014, 2015a, b) has three time scales. Dynamics of fuel cells evolves in at least three, possibly four, time scales

(Pukrushpan et al. 2004a; Zenith and Skogestad 2009). It was shown in Zenith and Skogestad (2009) that a proton exchange membrane fuel cell (PEMFC) system has three subsystems operating in three different time scales corresponding to three different time constants: electrochemical subsystem operating in seconds, chemical part of the PEMPC system (energy balance and mass balance) operating in minutes, and electrical part of the PEMFC system operating in milliseconds. Road vehicles display multi-time scale dynamics as shown in Wedig (2014). In power electronics many devices operate in three time scales (Umbria et al. 2014). Three time scales can be found in helicopters (Esteban et al. 2013).

Consider a three-time scale, time-invariant, linear dynamic control system represented by

$$\begin{bmatrix} \dfrac{dx_{\mathrm{I}}(t)}{dt} \\ \varepsilon_1 \dfrac{dx_{\mathrm{II}}(t)}{dt} \\ \varepsilon_2 \dfrac{dx_{\mathrm{III}}(t)}{dt} \end{bmatrix} = \begin{bmatrix} A_{11} & A_{12} & A_{13} \\ A_{21} & A_{22} & A_{23} \\ A_{31} & A_{32} & A_{33} \end{bmatrix} \begin{bmatrix} x_{\mathrm{I}}(t) \\ x_{\mathrm{II}}(t) \\ x_{\mathrm{III}}(t) \end{bmatrix} + \begin{bmatrix} B_{11} \\ B_{22} \\ B_{33} \end{bmatrix} u(t) \tag{4.28}$$

$$y(t) = \begin{bmatrix} C_{11} & C_{22} & C_{33} \end{bmatrix} \begin{bmatrix} x_{\mathrm{I}}(t) \\ x_{\mathrm{II}}(t) \\ x_{\mathrm{III}}(t) \end{bmatrix}$$

where $\varepsilon_1 \gg \varepsilon_2 > 0$ are small positive singular perturbation parameters, $x_{\mathrm{I}}(t) \in \mathrm{R}^{n_1}$ are slow state variables, $x_{\mathrm{II}}(t) \in \mathrm{R}^{n_2}$ are fast state variables, and $x_{\mathrm{III}}(t) \in \mathrm{R}^{n_3}$ are very fast state variables, $n = n_1+n_2+n_3$, $u(t) \in \mathrm{R}^m$ is the control input vector, $y(t) \in \mathrm{R}^p$ is the vector of system measurements, and A_{ij}, B_{ii}, and C_{ii},, $i,j = 1,2,3$, are constant matrices of appropriate dimensions. Matrices A_{11}, A_{22}, and A_{33} define, respectively, slow, fast, and very fast subsystems of dimensions n_1, n_2, and n_3 corresponding to state variables $x_{\mathrm{I}}(t)$, $x_{\mathrm{II}}(t)$, and $x_{\mathrm{III}}(t)$. It is a standard assumption in theory of three-time scale linear control systems that the matrices A_{22} and A_{33} are invertible (Kokotovic et al. 1999; Naidu and Calise 2001) so that the following assumption is imposed in the rest of this chapter.

Assumption 4.3 Matrices A_{22} and A_{33} are invertible.

The steps of the three-stage design used for three-time scale systems are obtained by modifying and simplifying the corresponding steps from Section 4.2 to the linear singularly perturbed control system whose state space form is defined in (4.28).

We start with the following change of variables applied to the original system defined in (4.28)

$$\eta_3(t) = L_1 x_{\mathrm{I}}(t) + L_2 x_{\mathrm{II}}(t) + x_{\mathrm{III}}(t) \tag{4.29}$$

which produces the dynamic equation for $\eta_3(t)$ as

$$\begin{aligned}\varepsilon_2 \frac{d\eta_3(t)}{dt} &= \left(A_{33} + \varepsilon_2 L_1 A_{13} + \frac{\varepsilon_2}{\varepsilon_1} L_2 A_{23}\right)\eta_3(t) + \left(B_{33} + \varepsilon_2 L_1 B_{11} + \frac{\varepsilon_2}{\varepsilon_1} L_2 B_{22}\right)u(t) \\ &\quad + f_{31}(L_1, L_2, \varepsilon_1, \varepsilon_2)x_{\mathrm{I}}(t) + f_{32}(L_1, L_2, \varepsilon_1, \varepsilon_2)x_{\mathrm{II}}(t) \\ &= A_3 \eta_3(t) + B_3 u(t) \\ A_3 &= A_{33} + \varepsilon_2 L_1 A_{13} + \frac{\varepsilon_2}{\varepsilon_1} L_2 A_{23}, \quad B_3 = B_{33} + \varepsilon_2 L_1 B_{11} + \frac{\varepsilon_2}{\varepsilon_1} L_2 B_{22}\end{aligned} \tag{4.30}$$

The corresponding algebraic equations for L_1 and L_2 matrices are given by

$$\begin{aligned} f_{31}(L_1, L_2, \varepsilon_1, \varepsilon_2) = A_{31} + \varepsilon_2 L_1 A_{11} + \frac{\varepsilon_2}{\varepsilon_1} L_2 A_{21} - \left(A_{33} + \varepsilon_2 L_1 A_{13} + \frac{\varepsilon_2}{\varepsilon_1} L_2 A_{23}\right) L_1 = 0 \\ f_{32}(L_1, L_2, \varepsilon_1, \varepsilon_2) = A_{32} + \varepsilon_2 L_1 A_{12} + \frac{\varepsilon_2}{\varepsilon_1} L_2 A_{22} - \left(A_{33} + \varepsilon_2 L_1 A_{13} + \frac{\varepsilon_2}{\varepsilon_1} L_2 A_{23}\right) L_2 = 0 \end{aligned} \tag{4.31}$$

Since $\varepsilon_1 \gg \varepsilon_2 \geq 0$ are small positive parameters, the above system of algebraic equations can be efficiently solved using the fixed-point iterations. Namely, by setting $\varepsilon_2 = 0$ in (4.31), we obtain the first-order approximations, which under Assumption 4.3 produce

$$\begin{aligned} A_{31} - A_{33} L_1^{(0)} = 0 \quad &\Rightarrow \quad L_1^{(0)} = A_{33}^{-1} A_{31} \\ A_{32} - A_{33} L_1^{(0)} = 0 \quad &\Rightarrow \quad L_2^{(0)} = A_{33}^{-1} A_{32} \end{aligned} \tag{4.32}$$

These approximate solutions are $O\left(\frac{\varepsilon_2}{\varepsilon_1}\right)$ close to the exact solutions, where the capitol O definition is as follows (Graham et al. 1989).

Definition 4.1 Quantities $O(\varepsilon)$, $O(\varepsilon^i)$, and $O(1)$ are defined, respectively, as

$$O(\varepsilon) < c\varepsilon, \quad O(\varepsilon^i) < c_i \varepsilon^i, \quad O(1) < k$$

where c, c_i, and k are bounded constants and i is an integer. This definition in the matrix sense should be interpreted as follows: A matrix is $O(\varepsilon^i)$ if the magnitude of its largest matrix element is $O(\varepsilon^i)$.

Using the capital O notation, we have

$$\begin{aligned} \left\| L_1 - L_1^{(0)} \right\| = O\left(\frac{\varepsilon_2}{\varepsilon_1}\right) \\ \left\| L_2 - L_2^{(0)} \right\| = O\left(\frac{\varepsilon_2}{\varepsilon_1}\right) \end{aligned} \tag{4.33}$$

It can be seen in (4.31) that the cross-coupling and nonlinear terms are multiplied either by ε_2 or $\varepsilon_2/\varepsilon_1$ so that the following fixed-point algorithm can be proposed for solving (4.31) in terms of *linear algebraic equations* starting with the initial conditions obtained in (4.32)

$$
\begin{aligned}
L_1^{(i+1)} &= A_{33}^{-1}\left[A_{31} + \varepsilon_2 L_1^{(i)} A_{11} + \frac{\varepsilon_2}{\varepsilon_1} L_2^{(i)} A_{21} - \left(\varepsilon_2 L_1^{(i)} A_{13} + \frac{\varepsilon_2}{\varepsilon_1} L_2^{(i)} A_{23}\right) L_1^{(i)}\right] \\
L_2^{(i+1)} &= A_{33}^{-1}\left[A_{32} + \varepsilon_2 L_1^{(i)} A_{12} + \frac{\varepsilon_2}{\varepsilon_1} L_2^{(i)} A_{22} - \left(\varepsilon_2 L_1^{(i)} A_{13} + \frac{\varepsilon_2}{\varepsilon_1} L_2^{(i)} A_{23}\right) L_2^{(i)}\right]
\end{aligned}
\tag{4.34}
$$

Running iterations, it is not difficult to show that this algorithm improves accuracy in each iteration by $O(\varepsilon_2/\varepsilon_1)$ so that after i iterations the following accuracy is obtained

$$
\begin{aligned}
\left\| L_1 - L_1^{(i)} \right\| &= O\left(\left(\frac{\varepsilon_2}{\varepsilon_1}\right)^i\right) \\
\left\| L_2 - L_2^{(i)} \right\| &= O\left(\left(\frac{\varepsilon_2}{\varepsilon_1}\right)^i\right)
\end{aligned}
\tag{4.35}
$$

Eliminating $x_{III}(t)$ from (4.28), the differential equations for $x_I(t)$ and $x_{II}(t)$ become

$$
\frac{dx_I(t)}{dt} = (A_{11} - A_{13}L_1)x_I(t) + (A_{12} - A_{13}L_2)x_{II}(t) + A_{13}\eta_3(t) + B_{11}u(t) \tag{4.36}
$$

and

$$
\varepsilon_1 \frac{dx_{II}(t)}{dt} = (A_{21} - A_{23}L_1)x_I(t) + (A_{22} - A_{23}L_2)x_{II}(t) + A_{23}\eta_3(t) + B_{22}u(t) \tag{4.37}
$$

Now, we introduce the second change of variables to remove dynamics of the first subsystem from the second subsystem

$$
\eta_2(t) = L_3 x_I(t) + x_{II}(t) \tag{4.38}
$$

This change of variables leads to

$$
\begin{aligned}
\varepsilon_1 \frac{d\eta_2(t)}{dt} &= [A_{22} - A_{23}L_2 + \varepsilon_1 L_3(A_{12} - A_{13}L_2)]\eta_2(t) \\
&\quad + (A_{23} + \varepsilon_1 L_3 A_{13})\eta_3(t) + f_{21}(L_1, L_2, L_3)x_I(t) \\
&\quad + (B_{22} + \varepsilon_1 L_3 B_{11})u(t) = A_2\eta_2(t) + (A_{23} + \varepsilon_1 L_3 A_{13})\eta_3(t) + B_2 u(t) \\
A_2 &= A_{22} - A_{23}L_2 + \varepsilon_1 L_3(A_{12} - A_{13}L_2), \quad B_2 = B_{22} + \varepsilon_1 L_3 B_{11}
\end{aligned}
\tag{4.39}
$$

where L_3 satisfies the following algebraic equation

$$(A_{21} - A_{23}L_1) - (A_{22} - A_{23}L_2)L_3 - \varepsilon_1 L_3(A_{12} - A_{13}L_2)L_3 + \varepsilon_1 L_3(A_{11} - A_{13}L_1) = 0 \tag{4.40}$$

In this nonlinear equation, the quadratic term is multiplied by ε_1 so that a fixed-point algorithm is efficient for solving (4.40) also. The zero-order approximation can be obtained from (4.40) by setting $\varepsilon_1 = 0$, which leads to

$$(A_{22} - A_{23}L_2)L_3^{(0)} = (A_{21} - A_{23}L_1) \quad \Rightarrow \quad L_3^{(0)} = (A_{22} - A_{23}L_2)^{-1}(A_{21} - A_{23}L_1) \tag{4.41}$$

The following assumption is needed in (4.41).

Assumption 4.4 The matrix $A_{22} - A_{23}L_2$ is invertible.

If this assumption is not satisfied, the initial condition can be obtained from the following algebraic Sylvester equation

$$(A_{22} - A_{23}L_2)L_3^{(0)} - \varepsilon_1 L_3^{(0)}(A_{11} - A_{13}L_1) = A_{21} - A_{23}L_1 \tag{4.42}$$

Since $\lambda(A_{22} - A_{23}L_2) = O(1)$ and $\lambda\{\varepsilon_1(A_{11} - A_{13}L_1)\} = O(\varepsilon_1)$, the coefficient matrices in (4.42) have no eigenvalues in common, and the unique solution of (4.42) exists (Chen 2012).

If Assumption 4.4 is satisfied, equation (4.40) can be solved iteratively using the fixed-point iterations as follows

$$\begin{aligned}(A_{22} - A_{23}L_2)L_3^{(i+1)} &= (A_{21} - A_{23}L_1) - \varepsilon_1 L_3^{(i)}(A_{12} - A_{13}L_2)L_3^{(i)} \\ &\quad + \varepsilon_1 L_3^{(i)}(A_{11} - A_{13}L_1)\end{aligned} \tag{4.43}$$

If Assumption 4.3 is not satisfied, algebraic equation (4.40) could be solved iteratively as an algebraic Sylvester equation using the following iterative algorithm:

$$\begin{aligned}&(A_{22} - A_{23}L_2)L_3^{(i+1)} - \varepsilon_1 L_3^{(i+1)}(A_{11} - A_{13}L_1) \\ &\quad = (A_{21} - A_{23}L_1) - \varepsilon_1 L_3^{(i)}(A_{12} - A_{13}L_2)L_3^{(i)}\end{aligned} \tag{4.44}$$

in which case an initial condition $L_3^{(0)}$ has to be chosen for (4.43), for example, by finding the least-square solution of the corresponding algebraic equation defined in (4.42). Note that the uniqueness of the solutions of L-equations is not required for the presented design methodology. It is not difficult to show that both algorithms converge with the rate of convergence of $O(\varepsilon_1)$ so that the accuracy of $O(\varepsilon_1^i)$ is achieved after i iterations, that is

$$\left\|L_3 - L_3^{(i)}\right\| = O\left(\varepsilon_1^i\right) \tag{4.45}$$

Transformation (4.41) applied to (4.31) produces the first subsystem in the new coordinates as

$$\begin{aligned}\frac{dx_{\mathrm{I}}(t)}{dt} &= [(A_{11} - A_{13}L_1) - (A_{12} - A_{13}L_2)L_3]x_{\mathrm{I}}(t) + (A_{12} - A_{13}L_2)\eta_2(t) \\ &\quad + A_{13}\eta_3(t) + B_{11}u(t) = A_{\mathrm{I}}x_{\mathrm{I}}(t) + (A_{12} - A_{13}L_2)\eta_2(t) + A_{13}\eta_3(t) + B_{11}u(t) \\ A_{\mathrm{I}} &= (A_{11} - A_{13}L_1) - (A_{12} - A_{13}L_2)L_3\end{aligned} \tag{4.46}$$

After the proposed transformation steps, the three-time scale singularly perturbed linear control system (4.28) is put in the upper block triangular form represented by differential equations (4.30), (4.39), and (4.46), whose state space form is given by

$$\begin{bmatrix} \dfrac{dx_{\mathrm{I}}(t)}{dt} \\ \varepsilon_1 \dfrac{d\eta_2(t)}{dt} \\ \varepsilon_2 \dfrac{d\eta_3(t)}{dt} \end{bmatrix} = \begin{bmatrix} A_{\mathrm{I}} & A_{12} - A_{13}L_2 & A_{13} \\ 0 & A_2 & A_{23} + \varepsilon_1 L_3 A_{13} \\ 0 & 0 & A_3 \end{bmatrix} \begin{bmatrix} x_{\mathrm{I}}(t) \\ \eta_2(t) \\ \eta_3(t) \end{bmatrix} + \begin{bmatrix} B_{11} \\ B_2 \\ B_3 \end{bmatrix} u(t) \tag{4.47}$$

The feedback controller design stages are presented in the following.

Stage 1
Apply feedback control

$$u(t) = -G_3\eta_3(t) + v(t)$$

to the η_3-subsystem, which leads to the form dual to (4.13), that is

$$\begin{bmatrix} \dfrac{dx_{\mathrm{I}}(t)}{dt} \\ \varepsilon_1 \dfrac{d\eta_2(t)}{dt} \\ \varepsilon_2 \dfrac{d\eta_3(t)}{dt} \end{bmatrix} = \begin{bmatrix} A_{\mathrm{I}} & A_{12} - A_{13}L_2 & A_{13} - B_{11}G_3 \\ 0 & A_2 & A_{23} + \varepsilon_1 L_3 A_{13} - B_2 G_3 \\ 0 & 0 & A_3 - B_3 G_3 \end{bmatrix} \begin{bmatrix} x_{\mathrm{I}}(t) \\ \eta_2(t) \\ \eta_3(t) \end{bmatrix} + \begin{bmatrix} B_{11} \\ B_2 \\ B_3 \end{bmatrix} v(t) \tag{4.48}$$

with the following matrices

$$
\begin{aligned}
A_1 &= (A_{11} - A_{13}L_1) - (A_{12} - A_{13}L_2)L_3 \\
A_2 &= A_{22} - A_{23}L_2 + \varepsilon_1 L_3(A_{12} - A_{13}L_2), \quad B_2 = B_{22} + \varepsilon_1 L_3 B_{11} \\
A_3 &= A_{33} + \varepsilon_2 L_1 A_{13} + \frac{\varepsilon_2}{\varepsilon_1} L_2 A_{23}, \quad B_3 = B_{33} + \varepsilon_2 L_1 B_{11} + \frac{\varepsilon_2}{\varepsilon_1} L_2 B_{22}
\end{aligned}
\tag{4.49}
$$

Stage 2

To isolate the second subsystem, we apply the change of variables as

$$
\xi_2(t) = \eta_2(t) - P_3\eta_3(t) \tag{4.50}
$$

which produces

$$
\begin{bmatrix} \dfrac{dx_1(t)}{dt} \\ \varepsilon_1 \dfrac{d\xi_2(t)}{dt} \\ \varepsilon_2 \dfrac{d\eta_3(t)}{dt} \end{bmatrix} = \begin{bmatrix} A_1 & A_{12} - A_{13}L_2 & A_{13} - B_{11}G_3 + (A_{12} - A_{13}L_2)P_3 \\ 0 & A_2 & 0 \\ 0 & 0 & A_3 - B_3G_3 \end{bmatrix} \begin{bmatrix} x_1(t) \\ \xi_2(t) \\ \eta_3(t) \end{bmatrix} + \begin{bmatrix} B_{11} \\ B_2 - \dfrac{\varepsilon_1}{\varepsilon_2} P_3 B_3 \\ B_3 \end{bmatrix} v(t) \tag{4.51}
$$

where P_3 satisfies the following algebraic equation

$$
\left(\frac{\varepsilon_2}{\varepsilon_1}\right) A_2 P_3 - P_3(A_3 - B_3G_3) + \left(\frac{\varepsilon_2}{\varepsilon_1}\right)(A_{23} + \varepsilon_1 L_3 A_{13} - B_2 G_3) = 0 \tag{4.52}
$$

Due to asymptotic stability of the very fast feedback matrix $(A_3 - B_3G_3)$, the solution for P_3 is unique and has the following features

$$
P_3 = O\left(\frac{\varepsilon_2}{\varepsilon_1}\right), \quad P_3^{(0)} = 0 \tag{4.53}
$$

The fixed-point algorithm for solving (4.52) up to the desired accuracy is given by

$$
P_3^{(i+1)}(A_3 - B_3G_3) = \left(\frac{\varepsilon_2}{\varepsilon_1}\right)\left(A_2 P_3^{(i)} + A_{23} + \varepsilon_1 L_3 A_{13} - B_2 G_3\right) \tag{4.54}
$$

with the initial condition defined in (4.53). It is not difficult to show that both algorithms converge with the rate of convergence of $O(\varepsilon_1)$ so that the accuracy of $O\left(\varepsilon_1^i\right)$ is achieved after i iterations, that is

$$\left\| P_3 - P_3^{(i)} \right\| = O\left(\frac{\varepsilon_2}{\varepsilon_1}\right) \tag{4.55}$$

In Stage 2, we apply feedback to control the second subsystem as

$$v(t) = -G_2\xi_2(t) + w(t)$$

which leads to

$$\begin{bmatrix} \frac{dx_1(t)}{dt} \\ \frac{\varepsilon_1 d\xi_2(t)}{dt} \\ \frac{\varepsilon_2 d\eta_3(t)}{dt} \end{bmatrix} = \begin{bmatrix} A_1 & A_{12} - A_{13}L_2 - B_{11}G_2 & A_{13} - B_{11}G_3 + (A_{12} - A_{13}L_2)P_3 \\ 0 & A_2 - \left(B_2 - \frac{\varepsilon_1}{\varepsilon_2}P_3B_3\right)G_2 & 0 \\ 0 & -B_3G_2 & A_3 - B_3G_3 \end{bmatrix}$$
$$\times \begin{bmatrix} x_1(t) \\ \xi_2(t) \\ \eta_3(t) \end{bmatrix} + \begin{bmatrix} B_{11} \\ B_2 - \frac{\varepsilon_1}{\varepsilon_2}P_3B_3 \\ B_3 \end{bmatrix} w(t) \tag{4.56}$$

Stage 3

Now we need to isolate the first subsystem and remove its coupling from subsystems 2 and 3. We first introduce a simplifying notation and present (4.56) in the form

$$\begin{bmatrix} \frac{dx_1(t)}{dt} \\ \varepsilon_1 \frac{d\xi_2(t)}{dt} \\ \varepsilon_2 \frac{d\eta_3(t)}{dt} \end{bmatrix} = \begin{bmatrix} \alpha_{11} & \alpha_{12} & \alpha_{13} \\ 0 & \alpha_{22} & 0 \\ 0 & \alpha_{32} & \alpha_{33} \end{bmatrix} \begin{bmatrix} x_1(t) \\ \xi_2(t) \\ \eta_3(t) \end{bmatrix} + \begin{bmatrix} \beta_1 \\ \beta_2 \\ \beta_3 \end{bmatrix} w(t) \tag{4.57}$$

with the obvious definitions of the newly introduced matrices. To achieve this goal, we apply another change of state variables as

$$\xi_1(t) = x_1(t) - P_1\xi_2(t) - P_2\eta_3(t) \tag{4.58}$$

This change of variables modifies (4.57) into

$$
\begin{bmatrix} \dfrac{d\xi_1(t)}{dt} \\ \varepsilon_1 \dfrac{d\xi_2(t)}{dt} \\ \varepsilon_2 \dfrac{d\eta_3(t)}{dt} \end{bmatrix} = \begin{bmatrix} \alpha_{11} & 0 & 0 \\ 0 & \alpha_{22} & 0 \\ 0 & \alpha_{32} & \alpha_{33} \end{bmatrix} \begin{bmatrix} \xi_1(t) \\ \xi_2(t) \\ \eta_3(t) \end{bmatrix} + \begin{bmatrix} \beta_1 - \dfrac{1}{\varepsilon_1} P_1 \beta_2 - \dfrac{1}{\varepsilon_2} P_2 \beta_3 \\ \beta_2 \\ \beta_3 \end{bmatrix} w(t) \tag{4.59}
$$

where P_1 and P_2 satisfy the following linear algebraic equations

$$
\frac{\varepsilon_2}{\varepsilon_1} P_1 \alpha_{22} - \varepsilon_2 \alpha_{11} P_1 + P_2 \alpha_{32} - \varepsilon_2 \alpha_{12} = 0 \tag{4.60}
$$

$$
\varepsilon_2 \alpha_{11} P_2 - P_2 \alpha_{33} + \varepsilon_2 \alpha_{13} = 0 \tag{4.61}
$$

Note that from these equations we have $P_2 = O(\varepsilon_2)$ and $P_1 = O(\varepsilon_1)$ so that the first subsystem input matrix in (4.59) is $O(1)$.

The solution for P_2 can be obtained from (4.61) either directly as a solution of the algebraic Sylvester equation or by performing the following fixed-point iterations

$$
P_2^{(i+1)} = \varepsilon_2 \left(\alpha_{11} P_2^{(i)} + \alpha_{13} \right) \alpha_{33}^{-1}, \quad P_2^{(0)} = 0, \quad i = 1,2, \ldots, N \tag{4.62}
$$

Having obtained the solution for P_2 from (4.62) after say N iterations, equation (4.60) can be solved either iteratively as

$$
P_1^{(i+1)} = \left[\varepsilon_1 \left(\alpha_{12} + \alpha_{11} P_1^{(i)} \right) - \frac{\varepsilon_1}{\varepsilon_2} P_2^{(N)} \alpha_{32} \right] \alpha_{22}^{-1}, \quad P_1^{(0)} = \frac{\varepsilon_1}{\varepsilon_2} P_2^{(N)} \alpha_{32} \alpha_{22}^{-1}, \quad i = 1,2, \ldots \tag{4.63}
$$

or directly as the Sylvester algebraic equation.

The local feedback control input can be applied to the first subsystem as $w(t) = -G_1 \xi_1(t)$, which leads to

$$\begin{bmatrix} \dfrac{d\xi_1(t)}{dt} \\ \varepsilon_1 \dfrac{d\xi_2(t)}{dt} \\ \varepsilon_2 \dfrac{d\eta_3(t)}{dt} \end{bmatrix} = \begin{bmatrix} \alpha_{11} - \left(\beta_1 - \dfrac{1}{\varepsilon_1} P_1 \beta_2 - \dfrac{1}{\varepsilon_2} P_2 \beta_3\right) G_1 & 0 & 0 \\ -\beta_2 G_1 & \alpha_{22} & 0 \\ -\beta_3 G_1 & \alpha_{32} & \alpha_{33} \end{bmatrix} \begin{bmatrix} \xi_1(t) \\ \xi_2(t) \\ \eta_3(t) \end{bmatrix} \tag{4.64}$$

It is important to emphasize that since $P_2 = O(\varepsilon_2)$ and $P_1 = O(\varepsilon_1)$, the first subsystem input matrix in (4.59) is $O(1)$ and the slow subsystem after feedback is applied remains slow.

The obtained system matrix is lower block triangular so its closed-loop eigenvalues are the union of the closed-loop eigenvalues of the subsystems. Of course, one can design different types of controllers for different subsystems not necessarily based on the eigenvalues assignment procedure. That will be demonstrated in the next section, where we will consider a real physical system, a proton exchange membrane fuel cell.

4.4 Application to a Proton Exchange Membrane Fuel Cell

All steps of the proposed three-stage design algorithm will be demonstrated in this section using a real physical system. The numerical values for the linearized three-time scale model of a proton exchange membrane (PEM) fuel cell are taken from Pukrushpan et al. (2004a, b). In the model considered in Pukrushpan et al. (2004a, b), on the anode side, hydrogen is provided from a tank via a supply manifold, and on the cathode side, air is pumped using a compressor. The corresponding state space matrices are given by

$$A = \begin{bmatrix} -6.30908 & 0 & -10.9544 & 0 & 83.74458 & 0 & 0 & 24.05866 \\ 0 & -161.083 & 0 & 0 & 51.52923 & 0 & -18.0261 & 0 \\ -18.7858 & 0 & -46.3136 & 0 & 275.6592 & 0 & 0 & 158.3741 \\ 0 & 0 & 0 & -17.3506 & 193.9373 & 0 & 0 & 0 \\ 1.299576 & 0 & 2.969317 & 0.3977 & -38.7024 & 0.105748 & 0 & 0 \\ 16.64244 & 0 & 38.02522 & 5.066579 & -479.384 & 0 & 0 & 0 \\ 0 & -450.386 & 0 & 0 & 142.2084 & 0 & -80.9472 & 0 \\ 2.02257 & 0 & 4.621237 & 0 & 0 & 0 & 0 & -51.2108 \end{bmatrix}$$

$$B = \begin{bmatrix} 0 & 0 & 0 & 3.946683 & 0 & 0 & 0 & 0 \end{bmatrix}^T$$

The state space variables represent the following quantities:

$$\begin{aligned} x(t) &= \begin{bmatrix} x_1(t) & x_2(t) & x_3(t) & x_4(t) & x_5(t) & x_6(t) & x_7(t) & x_8(t) \end{bmatrix}^T \\ &= \begin{bmatrix} m_{O_2}(t) & m_{H_2}(t) & m_{N_2}(t) & \omega_{cp}(t) & p_{sm}(t) & m_{sm}(t) & m_{H_2O_A}(t) & p_{rm}(t) \end{bmatrix} \end{aligned}$$

where $m_{O_2}, m_{H_2}, m_{N_2}, m_{H_2O_A}$, are, respectively, masses of oxygen, hydrogen, nitrogen, and anode side water vapor; ω_{cp} is the compressor (that blows the air (oxygen) on the cathode side) angular velocity; p_{sm} is the supply manifold gas pressure and m_{sp} is the mass of gas in the supply manifold; and p_{rm} is the return manifold gas pressure. $u(t) = v_{cm}(t)$ is the compressor motor voltage that represents the system input.

Finding the eigenvalues of the PEM fuel cell system matrix A, it can be seen that the eigenvalues are widely spread and that they can be clustered in three groups, three slow eigenvalues closest to the imaginary axis, three fast eigenvalues further from the imaginary axis, and two very fast eigenvalues far from the imaginary axis, that is

$$\begin{aligned}&\lambda_1 = -1.4038, \quad \lambda_2 = -1.6473, \quad \lambda_3 = -2.9151\\&\lambda_4 = -18.2582, \quad \lambda_5 = -22.4040, \quad \lambda_6 = -46.1768\\&\lambda_7 = -89.4853, \quad \lambda_8 = -219.6262\end{aligned}$$

The small singular perturbation parameters are determined either as ratios of the magnitudes of the eigenvalues from the corresponding clusters or ratios of the corresponding eigenvalue real parts, that is, in the case of the considered PEM fuel cell, they are given by

$$\varepsilon_1 = \frac{|\lambda_{slow\_\max}|}{|\lambda_{fast\_\min}|} = \frac{2.9151}{18.2582} = 0.1597,$$

$$\varepsilon_2 = \frac{|\lambda_{slow\_\max}|}{|\lambda_{veryfast\_\min}|} = \frac{2.9151}{89.4853} = 0.0326 \tag{4.65}$$

The PEM fuel cell mathematical model as defined in this section is in the *implicit* singularly perturbed form, namely, the system matrix A has the form $A = A(\varepsilon_1, \varepsilon_2)$. To get the *explicit* singularly form consistent with (4.28), we have to exchange the order of the state space variables such that matrices A_{22} and A_{33} corresponding to the fast and very fast variables are non-singular. In general, this is not an easy task, especially for higher-order dimensional systems. After many attempts, we have found that the exchange of state variables using a similarity matrix V obtained as a product of permutation matrices given by

$$V = I_{16} I_{25} I_{24} I_{27} I_{15} I_{36} \tag{4.66}$$

will produce the desired singularly perturbed form defined in (4.28). A permutation matrix I_{ij} is obtained from the identity matrix by interchanging rows i and j (Golub and Van Loan 2012). The matrix $A_{sp} = VAV^T$produces the explicit singularly structure for the considered PEM fuel cell mathematical model as

$$A_{sp} = VAV^T = \begin{bmatrix} A_{11sp} & A_{12sp} & A_{13sp} \\ A_{21sp} & A_{22sp} & A_{23sp} \\ A_{31sp} & A_{32sp} & A_{33sp} \end{bmatrix}$$

$$= \begin{bmatrix} -46.3136 & -18.7858 & 0 & 0 & 0 & 275.6592 & 0 & 158.3741 \\ -10.9544 & -6.3091 & 0 & 0 & 0 & 83.7446 & 0 & 24.0587 \\ 38.0252 & 16.6424 & 0 & 0 & 5.0666 & -479.3840 & 0 & 0 \\ 0 & 0 & 0 & -80.9472 & 0 & 142.2084 & -450.3860 & 0 \\ 0 & 0 & 0 & 0 & -17.3506 & 193.9373 & 0 & 0 \\ 2.9693 & 1.2996 & 0.1057 & 0 & 0.3977 & -38.7024 & 0 & 0 \\ 0 & 0 & 0 & -18.0261 & 0 & 51.5292 & -161.083 & 0 \\ 4.6212 & 2.0226 & 0 & 0 & 0 & 0 & 0 & -51.2108 \end{bmatrix}$$

The matrices in (4.28) are given by

$$\begin{aligned} A_{11} &= A_{11sp}, \quad A_{12} = A_{12sp}, \quad A_{13} = A_{13sp} \\ A_{21} &= \varepsilon_1 A_{21sp}, \quad A_{22} = \varepsilon_1 A_{22sp} \quad A_{23} = \varepsilon_1 A_{23sp} \\ A_{31} &= \varepsilon_2 A_{31sp}, \quad A_{32} = \varepsilon_2 A_{32sp} \quad A_{33} = \varepsilon_2 A_{33sp} \end{aligned} \tag{4.67}$$

What has been done for the original singularly perturbed linear control system in (4.66) and (4.67) is equivalent to ordering of the original state space variables in (4.28) as

$$[x_3(t) \;\; x_1(t) \;\; x_6(t) \;\; x_8(t) \;\; x_4(t) \;\; x_5(t) \;\; x_2(t) \;\; x_7(t)] \tag{4.68}$$

With this ordering of the state space variables, Assumption 4.3 is satisfied, the slow subsystem contains all slow eigenvalues, and the fast and very fast subsystems contain, respectively, all fast and all very fast eigenvalues. With this ordering of the state variables, we conclude that $x_3(t), x_1(t), x_6(t)$ (mass of nitrogen, mass of oxygen, mass of gas in the supply manifold) are slow state variables, $x_8(t), x_4(t), x_5(t)$ (gas pressure in the return manifold, compressor angular velocity, and supply manifold gas pressure) are fast state variables, and $x_2(t), x_7(t)$ (hydrogen mass and mass of water on the anode side) are very fast state variables.

Comment 4.2 It is interesting to observe the two most important PEM fuel cell variables: the pressure of oxygen (proportional to $x_1(t)$) is in the subsystem representing slow variables, and the pressure of hydrogen (proportional to $x_2(t)$) is in the subsystem representing very fast variables. Another interesting observation, even though expected from the physical point of view, can be verified mathematically by using the balancing transformation (Zhou and Doyle 1998) and finding the measures of the dynamic dominance of each state space variable in the considered singularly perturbed system. Namely, the balancing transformation reveals that the most dominant variables in this system (the variables with the highest energy) are $x_1(t)$ and $x_2(t)$. What is interesting here is the fact that $x_2(t)$ is dominant despite of being very fast (usually very fast state variables are low-energy signals and hence nondominant signals).

The input matrix in the explicit singularly perturbed form of the PEM fuel cell is

$$B_{sp} = \begin{bmatrix} B_{11sp} \\ B_{22sp} \\ B_{33sp} \end{bmatrix} = VB = \begin{bmatrix} 0 & 0 & 0 & 0 & 3.9467 & 0 & 0 & 0 \end{bmatrix}^T \tag{4.69}$$

The corresponding partitioned input matrices defined in (4.28) are given by

$$B_{11} = B_{11sp}, \quad B_{22} = \varepsilon_1 B_{22sp} \quad B_{33} = \varepsilon_2 B_{33sp} \tag{4.70}$$

It can be easily checked that the singularly perturbed system (4.28) represented by matrices A_{sp} and B_{sp} (considered in this section for the PEM fuel cell) is controllable (Chen 2012; Radisavljevic 2011), so that the corresponding subsystems defined in (4.47) by pairs (A_1, B_{11}), (A_2, B_2), and (A_3, B_3) are controllable also. Under the controllability condition, a controller can be designed to assign the desired subsystem eigenvalues (and hence the system eigenvalues since the eigenvalues are invariant under a similarity transformation, Chen (2012)) independently as

$$\lambda(A_1 - B_{11}G_1) = \lambda_{\mathrm{I}}^{\mathrm{desired}}, \quad \lambda(A_2 - B_2G_2) = \lambda_{\mathrm{II}}^{\mathrm{desired}}, \quad \lambda(A_3 - B_3G_3) = \lambda_{\mathrm{III}}^{\mathrm{desired}} \tag{4.71}$$

such that

$$\lambda(A - BG) = \lambda_{\mathrm{system}}^{\mathrm{desired}} = \left\{\lambda_{\mathrm{I}}^{\mathrm{desired}}, \ \lambda_{\mathrm{II}}^{\mathrm{desired}}, \ \lambda_{\mathrm{III}}^{\mathrm{desired}}\right\} \tag{4.72}$$

This procedure can be used also for partial assignment of eigenvalues via feedback. If, for example, one needs to change eigenvalues in the second subsystem, only then one sets $G_1 = 0, \ G_3 = 0$ and chooses G_2 to provide the desired eigenvalues for the second subsystem. In addition, if the subsystems (A_1, B_{11}), (A_2, B_2), and (A_3, B_3) are only stabilizable, the controller can be designed to stabilize only unstable subsystems and hence the original system. Moreover, different types of controllers can be designed for different subsystems. For example, we can design the eigenvalue assignment controller for the first subsystem and the linear-quadratic optimal controllers for the second and third subsystems.

Simulation results are presented for several cases.

Case 1 Eigenvalues assignment of all three subsystems.
The desired closed-loop eigenvalues are chosen as

$$\begin{aligned} &\lambda_1^{desired} = -2, \quad \lambda_{2,3}^{desired} = -3 \pm j2 \\ &\lambda_4^{desired} = -19, \quad \lambda_5^{desired} = -28, \quad \lambda_6^{desired} = -31 \\ &\lambda_7^{desired} = -100, \quad \lambda_8^{desired} = -150 \end{aligned}$$

The L- and P-equations were solved with accuracy of $O(10^{-12})$. The following local feedback gains are obtained using the eigenvalue assignment technique for the local reduced-order subsystems with numerically well-defined computational problems; the global full-order eigenvalue assignment problem for (4.28) is numerically ill-defined due to the singularly perturbed structures of the system matrix. Following the three-stage design procedure, the gains

$$G_3 = 10^4 \times [-2.0762 \quad -0.0013], \quad G_2 = [-149.4266 \quad -3.0934 \quad 505.1029]$$
$$G_1 = [160.3963 \quad 43.5540 \quad -115.2514]$$

set up the desired feedback eigenvalues with the accuracy of $O(10^{-12})$. The corresponding equivalent feedback gains in the original coordinates (obtained using (4.27)) are given by

$$G_{3eq} = 10^4 \times [-2.1343 \quad -0.0173], \quad G_{2eq} = 10^4 \times [-0.2911 \quad -0.0017 \quad 1.0046]$$
$$G_{1eq} = [-115.9525 \quad -74.8204 \quad -121.8927]$$

Case 2 The slow subsystem is optimized and no control for fast and very fast subsystems is needed.

In this case the fast and very fast subsystem gains are set to zero, that is,

$$G_3 = [0 \quad 0], \quad G_2 = [0 \quad 0 \quad 0]$$

and the slow subsystem gain is obtained via the linear-quadratic optimization of the isolated slow subsystem given in (4.59) with the quadratic performance criterion defined by

$$J_s = \frac{1}{2}\int_0^{\infty} \left(\xi_1^T(t)Q_s\xi_1(t) + w^T(t)R_s w(t)\right) dt\xi, \quad Q_s = I_3, \quad R_s = 1 \qquad (4.73)$$

For the reason of simplicity, we have assumed that the weighted matrices in this performance criterion are taken as identities. The corresponding optimal slow subsystem performance value is given by $J_s^{opt} = 1.1593$. The slow subsystem locally optimal feedback gain is obtained as

$$G_1^{opt} = [-0.0040 \quad 0.0094 \quad 0.1327]$$

The L- and P-equations in this case were solved with accuracy of $O(10^{-14})$, and the equivalent feedback gains in the original coordinates obtained via (4.27) are given by

$$G_{3eq} = [0 \quad 0.2102], \quad G_{2eq} = [0 \quad 0.0178 \quad -1.0745],$$
$$G_{1eq} = [0.0584 \quad 0.0362 \quad 0.1349]$$

These gains preserve the fast and very fast eigenvalues and produce the following feedback system eigenvalues (the slow ones are changed due to the impact of the slow optimal controller)

$$\lambda_1 = -1.3936, \quad \lambda_2 = -1.6552, \quad \lambda_3 = -2.9880$$
$$\lambda_4 = -18.2582, \quad \lambda_5 = -22.4040, \quad \lambda_6 = -46.1768$$
$$\lambda_7 = -89.4853, \quad \lambda_8 = -219.6262$$

Case 3 All three subsystems are optimized via the use of the local linear-quadratic optimal controllers.

In addition to optimizing the slow subsystem and the corresponding slow performance criterion (4.73), the fast and very fast feedback controllers are obtained by optimizing the corresponding performance criteria. The very fast subsystem is optimized in Step 1 (see (4.47)) using the following performance criterion

$$J_{vf} = \frac{1}{2}\int_0^\infty \left(\eta_3^T(t)Q_{vf}\eta_3(t) + u^T(t)R_{vf}u(t)\right) dt\xi, \quad Q_{vf} = I_2, \quad R_{vf} = 1 \qquad (4.74)$$

The fast subsystem is optimized in Step 2 (see (4.51)) for the quadratic performance criterion

$$J_f = \frac{1}{2}\int_0^\infty \left(\xi_2^T(t)Q_f\xi_2(t) + v^T(t)R_f v(t)\right)dt, \quad Q_f = I_3, \quad R_f = 1 \qquad (4.75)$$

The L- and P-equations were solved with accuracy of $O(10^{-14})$. The corresponding locally optimal gains are given by

$$G_3^{opt} = 10^{-2} \times [0.0063 \quad -0.1683], \quad G_2^{opt} = [0.0003 \quad 0.1068 \quad 0.1416]$$
$$G_1^{opt} = [-0.0040 \quad 0.0092 \quad 0.1292]$$

and the equivalent feedback gains (4.27) are

$$G_{3eq} = [-0.0007 \quad -0.1549], \quad G_{2eq} = [0.0002 \quad 0.1265 \quad -1.4104]$$
$$G_{1eq} = [-0.0243 \quad 0.0038 \quad 0.1330]$$

The corresponding locally optimal performance values are

$$J_{vf}^{opt} = 0.0079, \quad J_{f}^{opt} = 0.1319, \quad J_{s}^{opt} = 1.1600$$

4.5 Notes

Presentation of material in this chapter is based on the paper by Radisavljevic-Gajic et al. (2017). Permission for the use of such material in this research monograph was granted to us by the American Society of Mechanical Engineers (ASME) for the journal paper Radisavljevic-Gajic et al. (2017), published in *Transactions of ASME Journal of Dynamic Systems Measurement and Control.*

Chapter 5
Three-Stage Discrete-Time Feedback Controller Design

In this chapter, the results of two-stage feedback discrete-time controller design from Chap. 3 are extended to the three-stage feedback controller design of discrete-time, time-invariant, linear systems. In the case of a general discrete-time time-invariant linear system, the three-stage feedback controller design derivations practically parallel the derivations done for continuous-time, time-invariant, linear systems with the difference equations replacing the differential equations and arrive at the same sets of the three nonlinear algebraic equations and the three linear Sylvester algebraic equations. Those equations have to be solved in order to facilitate the considered three-stage feedback controller design. Consequently, assuming that the solutions of the corresponding linear and nonlinear algebraic equations are obtained, all good features of the three-stage feedback controller designed outlined for continuous-time linear systems in Chap. 4 hold in the case of discrete-time linear, time-invariant, systems presented in this chapter.

In the second part of the chapter, we discuss how to specialize the three-stage feedback controller design results obtained for general linear discrete-time time-invariant systems to three-time scale discrete-time linear, time-invariant, control systems (singularly perturbed discrete-time linear control systems) (Kokotovic et al. 1999; Naidu and Calise 2001; Gajic and Lim 2001; Dimitriev and Kurina 2006; Kuehn 2015). These systems have slow, fast, and very fast state space variables. It is well-known that many systems in mechanical and aerospace engineering possess singularly perturbed structure (Hsiao et al. 2001; Naidu and Calise 2001; Chen et al. 2002; Shapira and Ben-Asher 2004; Demetriou and Kazantzis 2005; Wang and Ghorbel 2006; Amjadifard et al. 2011; Kuehn 2015) due to the presence of small and large time constants, small masses, small moments of inertia, and small stiffness coefficients, which cause clustering of the system eigenvalues into two or several disjoint groups. The slower eigenvalue cluster is closer to the unit circle, and the faster eigenvalue cluster is closer to the origin.

V. Radisavljević-Gajić et al., *Multi-Stage and Multi-Time Scale Feedback Control of Linear Systems with Applications to Fuel Cells*, Mechanical Engineering Series,
https://doi.org/10.1007/978-3-030-10389-7_5

5.1 Three-Stage Discrete-Time Linear Feedback Controllers

Consider a time-invariant discrete-time linear dynamic system composed of three subsystems represented by the following difference equation:

$$\begin{bmatrix} x_{\mathrm{I}}(k+1) \\ x_{\mathrm{II}}(k+1) \\ x_{\mathrm{III}}(k+1) \end{bmatrix} = \begin{bmatrix} A_{11} & A_{12} & A_{13} \\ A_{21} & A_{22} & A_{23} \\ A_{31} & A_{32} & A_{33} \end{bmatrix} \begin{bmatrix} x_{\mathrm{I}}(k) \\ x_{\mathrm{II}}(k) \\ x_{\mathrm{III}}(k) \end{bmatrix} + \begin{bmatrix} B_{11} \\ B_{22} \\ B_{33} \end{bmatrix} u(k) = Ax(k) + Bu(k)$$
$$y(k) = [C_{11} \;\; C_{22} \;\; C_{33}] \begin{bmatrix} x_{\mathrm{I}}(k) \\ x_{\mathrm{II}}(k) \\ x_{\mathrm{III}}(k) \end{bmatrix} \tag{5.1}$$

where $k = 0,1,2,\ldots$, stands for discrete time; $x(k) \in \mathrm{R}^n$, $x_{\mathrm{I}}(k) \in \mathrm{R}^{n_1}$, $x_{\mathrm{II}}(t) \in \mathrm{R}^{n_2}$, and $x_{\mathrm{III}}(k) \in \mathrm{R}^{n_3}$, $n = n_1 + n_2 + n_3$, are discrete-time state space variables; $u(k) \in \mathrm{R}^m$ is the discrete-time system control input vector; $y(k) \in \mathrm{R}^p$ is the vector of discrete-time system measurements; and A_{ij}, B_{ii}, and C_{ii}, $i,j = 1,2,3$, are constant matrices of appropriate dimensions. Matrices A_{11}, A_{22}, and A_{33} define linear subsystems of dimensions n_1, n_2, and n_3, respectively, corresponding to the discrete-time state space variables $x_{\mathrm{I}}(k)$, $x_{\mathrm{II}}(k)$, and $x_{\mathrm{III}}(k)$. Matrices A_{ij}, i, $j = 1,2,3$, $i \neq j$ define couplings between the subsystems.

The steps of the three-stage feedback controller design for discrete-time linear systems parallel the steps of the corresponding design for continuous-time linear systems from Chap. 3. For the reason of completeness, and taking into account specific feature of discrete-time linear systems, these steps are completely presented in this section. Three time scales are present in several real physical systems. For example, advanced heavy water reactor (Shimjith et al. 2011a, b; Munje et al. 2014, 2016) has three time scales, with the last three papers using the discrete-time scale approach.

The discrete-time linear system is first mapped into appropriate coordinates via several transformations. We start with the following change of variables to the original discrete-time system defined in (5.1)

$$\eta_3(k) = L_1 x_{\mathrm{I}}(k) + L_2 x_{\mathrm{II}}(k) + x_{\mathrm{III}}(k) \tag{5.2}$$

which produces the linear difference equation for $\eta_3(k)$ as

$$\begin{aligned} \eta_3(k+1) = (A_{33} + L_1 A_{13} + L_2 A_{23})\eta_3(k) + (B_{33} + L_1 B_{11} + L_2 B_{22})u(k) \\ + f_{31}(L_1, L_2)x_I(k) + f_{32}(L_1, L_2)x_{II}(k) = A_3\eta_3(k) + B_3 u(k) \\ A_3 = A_{33} + L_1 A_{13} + L_2 A_{23}, \quad B_3 = B_{33} + L_1 B_{11} + L_2 B_{22} \end{aligned} \tag{5.3}$$

Elimination of coupling terms in (5.3) is achieved by assuming that the following system of algebraic equations has real solutions:

$$\begin{aligned} f_{31}(L_1, L_2) &= L_1A_{11} + L_2A_{21} + A_{31} - (L_1A_{13} + L_2A_{23} + A_{33})L_1 = 0 \\ f_{32}(L_1, L_2) &= L_1A_{12} + L_2A_{22} + A_{32} - (L_1A_{13} + L_2A_{23} + A_{33})L_2 = 0 \end{aligned} \tag{5.4}$$

Solving coupled nonlinear algebraic equations (5.4), in general, is not an easy task. It is expected that for the three-time scale discrete-time linear systems, the corresponding algebraic equations will have much simpler forms so that they potentially can be solved numerically using the fixed-point iterations as systems of linear algebraic equations.

Eliminating state variable $x_{\mathrm{III}}(k)$ from (5.1), the difference equations for $x_{\mathrm{I}}(k)$ and $x_{\mathrm{II}}(k)$ become

$$x_{\mathrm{I}}(k+1) = (A_{11} - A_{13}L_1)x_{\mathrm{I}}(k) + (A_{12} - A_{13}L_2)x_{\mathrm{II}}(k) + A_{13}\eta_3(k) + B_{11}u(k) \tag{5.5}$$

and

$$x_{\mathrm{II}}(k+1) = (A_{21} - A_{23}L_1)x_{\mathrm{I}}(k) + (A_{22} - A_{23}L_2)x_{\mathrm{II}}(k) + A_{23}\eta_3(k) + B_{22}u(k) \tag{5.6}$$

Since the $\eta_3(t)$-subsystem (5.3) is isolated, one can design a feedback controller for it by using $u(t) = -G_3\eta_3(t) + v(t)$, with $v(t)$ representing an input signal that can be used to control the first two subsystems. This strategy was used in the two-stage feedback controller design in Chap. 3. However, in this chapter due to a more complex three-stage design, we will first achieve an upper block triangular structure for the overall system and then show how to design independently local feedback controllers.

Now, we introduce the second change of variables to remove dynamics of the first subsystem from the second subsystem

$$\eta_2(k) = L_3x_{\mathrm{I}}(k) + x_{\mathrm{II}}(k) \tag{5.7}$$

This change of variables leads to

$$\begin{aligned} \eta_2(k+1) &= [A_{22} - A_{23}L_2 + L_3(A_{12} - A_{13}L_2)]\eta_2(k) \\ &\quad + (A_{23} + L_3A_{13})\eta_3(k) + f_{21}(L_1, L_2, L_3)x_{\mathrm{I}}(k) + (B_{22} + L_3B_{11})u(k) \\ &= A_2\eta_2(k) + (A_{23} + L_3A_{13})\eta_3(k) + B_2u(k) \\ &\quad A_2 = A_{22} - A_{23}L_2 + L_3(A_{12} - A_{13}L_2), \quad B_2 = B_{22} + L_3B_{11} \end{aligned} \tag{5.8}$$

where L_3 satisfies the following algebraic equation

$$L_3(A_{11}-A_{13}L_1)-(A_{22}-A_{23}L_2)L_3-L_3(A_{12}-A_{13}L_2)L_3+(A_{21}-A_{23}L_1)=0 \quad (5.9)$$

Assuming that L_1 and L_2 are previously obtained from (5.4), the matrix L_3 in fact satisfies the nonsymmetric, nonsquare algebraic Riccati equation that was previously studied in Medanic (1982) and Gao and Bai (2010). After transformation (5.7), the difference equation for the first subsystem in the new coordinates becomes

$$\begin{aligned} x_{\rm I}(k+1) &= [(A_{11}-A_{13}L_1)-(A_{12}-A_{13}L_2)L_3]x_{\rm I}(k) \\ &\quad +(A_{12}-A_{13}L_2)\eta_2(k)+A_{13}\eta_3(k)+B_{11}u(k) \\ &= A_1x_{\rm I}(k)+(A_{12}-A_{13}L_2)\eta_2(k)+A_{13}\eta_3(k)+B_{11}u(k) \\ A_1 &= (A_{11}-A_{13}L_1)-(A_{12}-A_{13}L_2)L_3 \end{aligned} \quad (5.10)$$

The transformations (5.2) and (5.7) relate the original and new state space variables as follows

$$\begin{bmatrix} x_{\rm I}(k) \\ \eta_2(k) \\ \eta_3(k) \end{bmatrix} = T_1 \begin{bmatrix} x_{\rm I}(k) \\ x_{\rm II}(k) \\ x_{\rm III}(k) \end{bmatrix} = \begin{bmatrix} I & 0 & 0 \\ L_3 & I & 0 \\ L_1 & L_2 & I \end{bmatrix} \begin{bmatrix} x_{\rm I}(k) \\ x_{\rm II}(k) \\ x_{\rm III}(k) \end{bmatrix} \quad (5.11a)$$

with the inverse transformation given by

$$\begin{bmatrix} x_{\rm I}(k) \\ x_{\rm II}(k) \\ x_{\rm III}(k) \end{bmatrix} = T_1^{-1} \begin{bmatrix} x_{\rm I}(k) \\ \eta_2(k) \\ \eta_3(k) \end{bmatrix} = \begin{bmatrix} I & 0 & 0 \\ -L_3 & I & 0 \\ -L_1+L_2L_3 & -L_2 & I \end{bmatrix} \begin{bmatrix} x_{\rm I}(k) \\ \eta_2(k) \\ \eta_3(k) \end{bmatrix} \quad (5.11b)$$

This similarity transformation maps the original system (5.1) into the upper block triangular form represented by difference equations (5.3), (5.8), and (5.10), whose state space form is given by

$$\begin{gathered} \begin{bmatrix} x_{\rm I}(k+1) \\ \eta_2(k+1) \\ \eta_3(k+1) \end{bmatrix} = \begin{bmatrix} A_1 & A_{12}-A_{13}L_2 & A_{13} \\ 0 & A_2 & A_{23}+L_3A_{13} \\ 0 & 0 & A_3 \end{bmatrix} \begin{bmatrix} x_{\rm I}(k) \\ \eta_2(k) \\ \eta_3(k) \end{bmatrix} + \begin{bmatrix} B_{11} \\ B_2 \\ B_3 \end{bmatrix} u(k) = \bar{A} \begin{bmatrix} x_{\rm I}(k) \\ \eta_2(k) \\ \eta_3(k) \end{bmatrix} + \bar{B}u(k) \\ A_1=(A_{11}-A_{13}L_1)-(A_{12}-A_{13}L_2)L_3 \\ A_2=A_{22}-A_{23}L_2+L_3(A_{12}-A_{13}L_2), \ B_2=B_{22}+L_3B_{11} \\ A_3=A_{33}+L_1A_{13}+L_2A_{23}, \ B_3=B_{33}+L_1B_{11}+L_2B_{22} \end{gathered} \quad (5.12)$$

$$y(k) = (C_{11} - C_{22}L_3 - C_{33}L_1)x_1(k) + (C_{22} - C_{33}L_2)\eta_2(k) + C_{33}\eta_3(k)$$
$$= C_1x_1(k) + C_2\eta_2(k) + C_{33}\eta_3(k) = \bar{C}\begin{bmatrix} x_1(k) \\ \eta_2(k) \\ \eta_3(k) \end{bmatrix}$$
$$C_1 = C_{11} - C_{22}L_3 - C_{33}L_1 + C_{33}L_2L_3, \quad C_2 = C_{22} - C_{33}L_2$$

Having transformed the original discrete-time linear system into an upper triangular form, we can start the design of discrete-time linear feedback controllers. We will indicate intermediate steps needed to make these designs independent. If one plans to design *optimal* linear-quadratic feedback controllers, the transformed system must be controllable-observable or at least stabilizable-detectable. Hence, the following assumption is needed.

Assumption 5.1 The considered discrete-time linear system (5.1) is controllable (stabilizable) and observable (detectable).

In the next remark, we clarify this issue by imposing these conditions on the original system (5.1) and indicate that they hold for the transformed system (5.12) as well.

Remark 5.1 If the original system (5.1) is controllable and observable, then the transformed system (5.12) is also controllable and observable since the original and new coordinates are related via the similarity transformation (5.11), which is the very well-known result that states that the controllability and observability rank tests are invariant under similarity transformations; see, for example, Chen (2012) and Sinha (2007). The same invariance result was established in Remark 4.1 for the stabilizability and detectability conditions, namely, it was shown in Remark 4.1 that if the original system (5.1) is stabilizable and detectable, then the transformed system (5.12) is also stabilizable and detectable. The *stabilizability-detectability invariance* result established in Remark 4.1 was obtained using the Popov-Belevitch-Hautus eigenvector stabilizability test (Zhou and Doyle 1998).

Now we start the three-stage discrete-time feedback controller design procedure, in which we isolate every discrete-time subsystem and identify the control input that independently of other subsystems solely controls the isolated subsystem.

Stage 1
Apply feedback control

$$u(k) = -G_3\eta_3(k) + v(k)$$

to the η-subsystem, which leads to the new linear dynamic system described by the following difference equation

$$\begin{bmatrix} x_1(k+1) \\ \eta_2(k+1) \\ \eta_3(k+1) \end{bmatrix} = \begin{bmatrix} A_1 & A_{12} - A_{13}L_2 & A_{13} - B_{11}G_3 \\ 0 & A_2 & A_{23} + L_3A_{13} - B_2G_3 \\ 0 & 0 & A_3 - B_3G_3 \end{bmatrix} \begin{bmatrix} x_1(k) \\ \eta_2(k) \\ \eta_3(k) \end{bmatrix} + \begin{bmatrix} B_{11} \\ B_2 \\ B_3 \end{bmatrix} v(k) \tag{5.13}$$

Stage 2

To isolate the second discrete-time subsystem, we apply the change of variables as

$$\xi_2(k) = \eta_2(k) - P_3\eta_3(k) \tag{5.14}$$

which produces the following linear difference equation

$$\begin{bmatrix} x_1(k+1) \\ \xi_2(k+1) \\ \eta_3(k+1) \end{bmatrix} = \begin{bmatrix} A_1 & A_{12} - A_{13}L_2 & A_{13} - B_{11}G_3 + (A_{12} - A_{13}L_2)P_3 \\ 0 & A_2 & 0 \\ 0 & 0 & A_3 - B_3G_3 \end{bmatrix} \begin{bmatrix} x_1(k) \\ \xi_2(k) \\ \eta_3(k) \end{bmatrix} + \begin{bmatrix} B_{11} \\ B_2 - P_3B_3 \\ B_3 \end{bmatrix} v(k) \tag{5.15}$$

where P_3 satisfies the following algebraic equation

$$A_2P_3 - P_3(A_3 - B_3G_3) + A_{23} + L_3A_{13} - B_2G_3 = 0 \tag{5.16}$$

Linear matrix algebraic equation (5.16) is the Sylvester algebraic equation. The existence of its unique solution is guaranteed under condition that matrices A_2 and $A_3 - B_3G_3$ have no eigenvalues in common (Chen 2012). Hence, the following assumption is needed.

Assumption 5.2 Matrices A_2 and $A_3 - B_3G_3$ have no eigenvalues in common.

This assumption is easily satisfied since $A_3 - B_3G_3$ is the feedback matrix of the third discrete-time subsystem, and the linear algebraic equation (5.16) has a unique solution.

In Stage 2, we apply feedback control to the second discrete-time subsystem as

$$v(k) = -G_2\xi_2(k) + w(k)$$

which leads to the following linear difference equation

$$\begin{bmatrix} x_1(k+1) \\ \xi_2(k+1) \\ \eta_3(k+1) \end{bmatrix} = \begin{bmatrix} A_1 & A_{12}-A_{13}L_2-B_{11}G_2 & A_{13}-B_{11}G_3+(A_{12}-A_{13}L_2)P_3 \\ 0 & A_2-(B_2-P_3B_3)G_2 & 0 \\ 0 & -B_3G_2 & A_3-B_3G_3 \end{bmatrix} \begin{bmatrix} x_1(k) \\ \xi_2(k) \\ \eta_3(k) \end{bmatrix} + \begin{bmatrix} B_{11} \\ B_2-P_3B_3 \\ B_3 \end{bmatrix} w(k) \tag{5.17}$$

Stage 3

Now we need to isolate the first discrete-time subsystem and remove its coupling from discrete-time subsystems 2 and 3. Before we proceed with this task, we first introduce the simplifying notation and present (5.17) in the form

$$\begin{bmatrix} x_1(k+1) \\ \xi_2(k+1) \\ \eta_3(k+1) \end{bmatrix} = \begin{bmatrix} \alpha_{11} & \alpha_{12} & \alpha_{13} \\ 0 & \alpha_{22} & 0 \\ 0 & \alpha_{32} & \alpha_{33} \end{bmatrix} \begin{bmatrix} x_1(k) \\ \xi_2(k) \\ \eta_3(k) \end{bmatrix} + \begin{bmatrix} \beta_1 \\ \beta_2 \\ \beta_3 \end{bmatrix} w(k) \tag{5.18}$$

with the obvious definitions of the newly introduced matrices. To isolate the first discrete-time subsystem, we apply another change of state variables defined as follows

$$\xi_1(k) = x_1(k) - P_1\xi_2(k) - P_2\eta_3(k) \tag{5.19}$$

This change of variables modifies (5.18) into

$$\begin{bmatrix} \xi_1(k+1) \\ \xi_2(k+1) \\ \eta_3(k+1) \end{bmatrix} = \begin{bmatrix} \alpha_{11} & 0 & 0 \\ 0 & \alpha_{22} & 0 \\ 0 & \alpha_{32} & \alpha_{33} \end{bmatrix} \begin{bmatrix} \xi_1(k) \\ \xi_2(k) \\ \eta_3(k) \end{bmatrix} + \begin{bmatrix} \beta_1 - P_1\beta_2 - P_2\beta_3 \\ \beta_2 \\ \beta_3 \end{bmatrix} w(k) \tag{5.20}$$

where P_1 and P_2 satisfy the following system of two linear matrix algebraic equations

$$\alpha_{11}P_1 - P_1\alpha_{22} - P_2\alpha_{32} + \alpha_{12} = 0 \tag{5.21}$$

$$\alpha_{11}P_2 - P_2\alpha_{33} + \alpha_{13} = 0 \tag{5.22}$$

The solution for P_2 can be directly obtained from (5.22) as a solution of the algebraic Sylvester equation. Having obtained the solution for P_2, equation (5.21) can be solved as another Sylvester equation that directly produces P_1. The unique solutions of these Sylvester equations exist under the following assumptions.

Assumption 5.3a (needed for (5.22)) Matrices α_{11} and α_{33} have no eigenvalues in common.

Assumption 5.3b (needed for (5.21)) Matrices α_{11} and α_{22} have no eigenvalues in common.

Since α_{22} and α_{33} are feedback matrices for the second and the third discrete-time subsystems, both assumptions can be easily satisfied.

Now the local feedback control input with the partial state feedback coming only from the first subsystem can be applied to the first subsystem as

$$w(k) = -G_1\xi_1(k)$$

which leads to the following closed-loop linear discrete-time system:

$$\begin{bmatrix} \xi_1(k+1) \\ \xi_2(k+1) \\ \eta_3(k+1) \end{bmatrix} = \begin{bmatrix} \alpha_{11} - (\beta_1 - P_1\beta_2 - P_2\beta_3)G_1 & 0 & 0 \\ -\beta_2 G_1 & \alpha_{22} & 0 \\ -\beta_3 G_1 & \alpha_{32} & \alpha_{33} \end{bmatrix} \begin{bmatrix} \xi_1(k) \\ \xi_2(k) \\ \eta_3(k) \end{bmatrix} \tag{5.23}$$

The obtained system matrix is block lower triangular so its closed-loop eigenvalues are the union of the closed-loop eigenvalues of the individual subsystems (diagonal blocks).

We can now relate the new and original state variables via a similarity transformation. The state coordinates defined in (5.12) and (5.23) are related via state transformations (5.14) and (5.19). These relationships can be put in a compact matrix form providing a transformation to the original state variables as follows

$$\begin{aligned} \begin{bmatrix} \xi_1(k) \\ \xi_2(k) \\ \eta_3(k) \end{bmatrix} &= \begin{bmatrix} I & -P_1 & -P_2 \\ 0 & I & 0 \\ 0 & 0 & I \end{bmatrix} \begin{bmatrix} x_{\mathrm{I}}(k) \\ \xi_2(k) \\ \eta_3(k) \end{bmatrix} = T_3 \begin{bmatrix} x_{\mathrm{I}}(k) \\ \xi_2(k) \\ \eta_3(k) \end{bmatrix} = T_3 \begin{bmatrix} I & 0 & 0 \\ 0 & I & -P_3 \\ 0 & 0 & I \end{bmatrix} \begin{bmatrix} x_{\mathrm{I}}(k) \\ \eta_2(k) \\ \eta_3(k) \end{bmatrix} \\ &= T_3T_2 \begin{bmatrix} x_{\mathrm{I}}(k) \\ \eta_2(k) \\ \eta_3(k) \end{bmatrix} = T_3T_2 \begin{bmatrix} I & 0 & 0 \\ L_3 & I & 0 \\ L_1 & L_2 & I \end{bmatrix} \begin{bmatrix} x_{\mathrm{I}}(k) \\ x_{\mathrm{II}}(k) \\ x_{\mathrm{III}}(k) \end{bmatrix} = T_3T_2T_1 \begin{bmatrix} x_{\mathrm{I}}(k) \\ x_{\mathrm{II}}(k) \\ x_{\mathrm{III}}(k) \end{bmatrix} = T \begin{bmatrix} x_{\mathrm{I}}(k) \\ x_{\mathrm{II}}(k) \\ x_{\mathrm{III}}(k) \end{bmatrix} \\ &= \begin{bmatrix} I - P_1L_3 - P_2L_1 + P_1P_3L_1 & -P_1 - P_2L_2 + P_1P_3L_2 & P_1P_3 - P_2 \\ L_3 - P_3L_1 & I - P_3L_2 & -P_3 \\ L_1 & L_2 & I \end{bmatrix} \begin{bmatrix} x_{\mathrm{I}}(k) \\ x_{\mathrm{II}}(k) \\ x_{\mathrm{III}}(k) \end{bmatrix} \end{aligned} \tag{5.24}$$

The discrete-time feedback control signal applied to the system in the transformed coordinates is given by

$$
\begin{aligned}
u(\xi_1(k), \xi_2(k), \eta_3(k)) &= -G_1\xi_1(k) - G_2\xi_2(k) - G_3\eta_3(k) \\
&= -\begin{bmatrix} G_1 & G_2 & G_3 \end{bmatrix} \begin{bmatrix} \xi_1(k) \\ \xi_2(k) \\ \eta_3(k) \end{bmatrix} \\
&= -\begin{bmatrix} G_1 & G_2 & G_3 \end{bmatrix} T \begin{bmatrix} x_{\rm I}(k) \\ x_{\rm II}(k) \\ x_{\rm III}(k) \end{bmatrix}
\end{aligned}
\tag{5.25}
$$

Using the state transformation (5.24), we can obtain the discrete-time feedback control signal in the original coordinates

$$
u(x_{\rm I}(k), x_{\rm II}(k), x_{\rm III}(k)) = -G_{1eq}x_{\rm I}(k) - G_{2eq}x_{\rm II}(k) - G_{3eq}x_{\rm III}(k) \tag{5.26}
$$

with the equivalent feedback gains given by

$$
\begin{aligned}
G_{1eq} &= G_1(I - P_1L_3 - P_2L_1 + P_1P_3L_1) + G_2(L_3 - P_3L_1) + G_3L_1 \\
G_{2eq} &= G_1(P_1P_3L_2 - P_1 - P_2L_2) + G_2(I - P_3L_2) + G_3L_2 \\
G_{3eq} &= G_1(P_1P_3 - P_2) - G_2P_3 + G_3
\end{aligned}
\tag{5.27}
$$

The three-stage discrete-time controller design get simplified in the case of three-time scale linear systems that have a natural decomposition into slow, fast, and very fast subsystems. Namely, the corresponding design equations get simpler forms, and due to the presence of two small parameters (that identify the time scales), the L- and P-equations can be easily solved.

5.2 Three-Stage Three-Time Scale Discrete Linear Control Systems

Discrete-time linear control systems with slow and fast modes are particularly well suited for the considered three-stage feedback controller design. For this class of systems, in general, numerical ill-conditioning appears if one attempts to design a linear feedback controller directly using the entire (full-order) system. Design simplifications can be achieved potentially by specializing the presented design from Sect. 5.1 to singularly perturbed linear discrete-time systems operating in three time scales.

Several multi-time scales are present in real physical systems that have components of different nature. For example, advanced heavy water reactor (Smimjith et al. 2011a, b; Munje et al. 2014, 2016) has three time scales. Dynamics of fuel cells evolves in at least three, possibly four, time scales (Pukrushpan et al. 2004a; Zenith and Skogestad 2009). It was shown in Zenith and Skogestad (2009) that a proton exchange membrane fuel cell (PEMFC) system has three subsystems operating in three different time scales corresponding to three different time constants: electrochemical subsystem operating in seconds, chemical part of the PEMPC system (energy balance and mass balance) operating in minutes, and electrical part of the PEMFC system operating in milliseconds. Road vehicles display multi-time scale dynamics as shown in Wedig (2014). In power electronics many devices operate in three time scales (Umbria et al. 2014). Three time scales can be found in helicopters (Esteban et al. 2013).

The issue about the most convenient three-time scale linear discrete-time system formulation has not been settled down yet, despite of the fact that the two-time scale discrete-time systems were formulated and fully understood in the 1980s. Mahmoud (1986), studying stabilization problem of three-time scale discrete-time linear systems, extended the two-time scale fast-time scale formulation of Litkouhi and Khalil (1984, 1985) and derived the following expression for such a control system

$$\begin{bmatrix} x_{\mathrm{I}}(k+1) \\ x_{\mathrm{II}}(k+1) \\ x_{\mathrm{III}}(k+1) \end{bmatrix} = \begin{bmatrix} I+\varepsilon_1 A_{11} & \varepsilon_1 A_{12} & \varepsilon_1 A_{13} \\ \frac{\varepsilon_1}{\varepsilon_2} A_{21} & \frac{\varepsilon_1}{\varepsilon_2} A_{22} & \frac{\varepsilon_1}{\varepsilon_2} A_{23} \\ \frac{\varepsilon_1}{\varepsilon_3} A_{31} & \frac{\varepsilon_1}{\varepsilon_3} A_{32} & \frac{\varepsilon_1}{\varepsilon_3} A_{33} \end{bmatrix} \begin{bmatrix} x_{\mathrm{I}}(k) \\ x_{\mathrm{II}}(t) \\ x_{\mathrm{III}}(t) \end{bmatrix} + \begin{bmatrix} \varepsilon_1 B_{11} \\ \frac{\varepsilon_1}{\varepsilon_2} B_{22} \\ \frac{\varepsilon_1}{\varepsilon_3} B_{33} \end{bmatrix} u(k) \tag{5.28}$$

$x_{\mathrm{I}}(k) \in \mathrm{R}^{n_1}$ are slow state variables, $x_{\mathrm{II}}(k) \in \mathrm{R}^{n_2}$ are fast state variables, and $x_{\mathrm{III}}(k) \in \mathrm{R}^{n_3}$ are very fast state variables, $n = n_1 + n_2 + n_3$, $u(k) \in \mathrm{R}^m$ is the control input vector, $y(k) \in \mathrm{R}^p$ is the vector of system measurements, and A_{ij}, B_{ii}, $i, j = 1,2,3$ are constant matrices of appropriate dimensions. Matrices A_{11}, A_{22}, and A_{33} define, respectively, slow, fast, and very fast subsystems of dimensions n_1, n_2, and n_3 corresponding to state variables $x_{\mathrm{I}}(k)$, $x_{\mathrm{II}}(k)$, and $x_{\mathrm{III}}(k)$. According to Mahmoud (1986), the small singular perturbation parameters satisfy $\varepsilon_1 = \sqrt{\varepsilon_2^2 + \varepsilon_3^2}$. Another choice for ε_1 could have been taken as $\varepsilon_1 = \sqrt{\varepsilon_2 \varepsilon_3}$ (Kokotovic et al. 1999).

An extension to discrete three-time scales of the fast-time scale formulation of Litkouhi and Khalil (1984, 1985) that appears to be more directly align with their original two-time scale formulation than the one presented in Mahmoud (1986) can be defined as follows:

$$\begin{bmatrix} x_{\mathrm{I}}(k+1) \\ x_{\mathrm{II}}(k+1) \\ x_{\mathrm{III}}(k+1) \end{bmatrix} = \begin{bmatrix} I+\varepsilon_1 A_{11} & \varepsilon_1 A_{12} & \varepsilon_1 A_{13} \\ \varepsilon_2 A_{21} & I+\varepsilon_2 A_{22} & \varepsilon_2 A_{23} \\ A_{31} & A_{32} & A_{33} \end{bmatrix} \begin{bmatrix} x_{\mathrm{I}}(k) \\ x_{\mathrm{II}}(t) \\ x_{\mathrm{III}}(t) \end{bmatrix} + \begin{bmatrix} \varepsilon_1 B_{11} \\ \varepsilon_2 B_{22} \\ B_{33} \end{bmatrix} u(k) \tag{5.29}$$

It is interesting to point out that studying the stability of three-time scale singularly perturbed discrete-time linear systems (Zerizer 2016) came up with the following system formulation:

$$\begin{bmatrix} x_{\mathrm{I}}(k+1) \\ x_{\mathrm{II}}(k+1) \\ \varepsilon_1 x_{\mathrm{III}}(k+1) \end{bmatrix} = \begin{bmatrix} A_{11} & \varepsilon_1 A_{12} & A_{13} \\ A_{21} & \varepsilon_1 A_{22} & A_{23} \\ A_{31} & \varepsilon_1 A_{32} & A_{33} \end{bmatrix} \begin{bmatrix} x_{\mathrm{I}}(k) \\ x_{\mathrm{II}}(t) \\ x_{\mathrm{III}}(t) \end{bmatrix} \tag{5.30}$$

Since no control problem was considered in Zerizer (2016), the input matrix B was not defined in that paper. It is interesting to observe that formulation of three-time scale linear discrete system (5.30) has only one singular perturbation parameter.

We formally may extend the two-time slow-time scale formulation of Rao and Naidu (1981) and Naidu and Calise (2001), which leads to the following linear discrete-time control system problem formulation with three time scales:

$$\begin{bmatrix} x_{\mathrm{I}}(k+1) \\ x_{\mathrm{II}}(k+1) \\ x_{\mathrm{III}}(k+1) \end{bmatrix} = \begin{bmatrix} A_{11} & \varepsilon_1 A_{12} & \varepsilon_2 A_{13} \\ A_{21} & \varepsilon_1 A_{22} & \varepsilon_2 A_{23} \\ A_{31} & \varepsilon_1 A_{32} & \varepsilon_2 A_{33} \end{bmatrix} \begin{bmatrix} x_{\mathrm{I}}(k) \\ x_{\mathrm{II}}(t) \\ x_{\mathrm{III}}(t) \end{bmatrix} + \begin{bmatrix} B_{11} \\ B_{22} \\ B_{33} \end{bmatrix} u(k) \tag{5.31}$$

Another problem formulation of discrete-time three-time scale linear systems can be found in the work of Munje et al. (2015a, b), where they in general assume that such systems have the eigenvalues clustered in three disjoint rings inside of the unit circle. Elements of such a formulation can be found in the classic works of Philipps (1980a, 1983) and Naidu (1988). In those works, the small singular perturbation parameters were not introduced, so this problem formulation will not be beneficial in simplifying the derived algebraic equations of Sect. 5.2 of the three-stage feedback controller design. Those algebraic equations can be simplified only if the singular perturbation parameters are identified and their order of magnitudes clearly determined.

5.3 Future Research Topics

It is necessary first to justify which of the presented four three-time scale formulations of discrete-time linear systems provides the best agreement with the actual discrete-time system such that the design and system analysis can be completely performed at the subsystem levels using meaningful control-oriented assumptions and conditions. For such obtained formulation, one should study numerical techniques for solving the design nonlinear algebraic equations defined in (5.4), (5.9), and linear Sylvester algebraic equations defined in (5.16), (5.21), and (5.22), hopefully all of them in terms of linear algebraic equations by developing corresponding fixed-point iterations (or Newton method type) algorithms. To that end, the presence of small singular perturbation parameters will play an important role.

Chapter 6
Four-Stage Continuous-Time Feedback Controller Design

In this chapter, the results of three-stage continuous-time feedback controller design from Chap. 4 are extended to the four-stage feedback controller design. This facilitates independent control of four subsets of system state variables representing four artificial or natural subsystems of a system under consideration. The newly derived technique requires only solutions of reduced-order subsystem level algebraic equations for the design of appropriate local feedback controllers using only the corresponding subsystem state feedback (partial feedback). The local feedback controllers are combined to form a global controller for the system under consideration. The technique presented facilitates designs of independent full-state feedback controllers at the subsystem levels. Different types of local controllers, for example, eigenvalue assignment, robust, optimal in some sense, and observer-based controllers, may be used to control different subsystems. This feature has not been available for any other known linear feedback controller design technique.

In the second part of the chapter, we indicate how to specialize the results obtained to the four time-scale linear control systems (singularly perturbed control systems) that have natural decomposition into four time scales: very slow, slow, fast, and very fast subsystems. Moreover, the proposed technique eliminates numerical ill-conditioning of the original four-time scale singularly perturbed linear systems.

6.1 Introduction

The technique to be presented requires that the system subsystems are clearly identified. The system partitioning into four subsystems can be done using several methods: (1) based on the physical nature of subsystems (system natural decomposition), (2) according to the conditions that must be satisfied such that the partitioned system is feasible for the four-stage feedback design, (3) based on mathematical conditions that must be satisfied to solve the corresponding design equations,

V. Radisavljević-Gajić et al., *Multi-Stage and Multi-Time Scale Feedback Control of Linear Systems with Applications to Fuel Cells*, Mechanical Engineering Series,
https://doi.org/10.1007/978-3-030-10389-7_6

(4) control needs (which parts of the system should be independently controlled via best suited local controllers), and/or (5) grouping of the state space variables such that subsystems satisfy control-oriented assumptions, for example, needed for the design of local optimal controllers and filters that require controllability (stabilizability) and or observability (detectability) conditions. It is possible that the system stabilizability-detectability conditions are not satisfied so that the system optimal controller cannot be designed, but the stabilizability-detectability conditions can be satisfied for local subsystems and locally optimal controllers can be designed.

The power of the four-stage feedback design is in the following:

(a) Different types of controllers (optimal, eigenvalue assignment, robust, reliable, etc.) can be designed for different parts of the system (subsystems) using corresponding feedback gains obtained by performing calculations (design) only with subsystem (reduced-order) matrices.
(b) Local subsystem feedback gains that control local subsystems are compounded into one full-state feedback gain via a single formula, leading to a unified feedback controller for the system under consideration.
(c) Computational requirements are drastically reduced since computations are done with matrices of reduced dimensions corresponding to subsystems.
(d) Very high accuracy can be achieved since numerical ill-conditioning of higher-order matrices can be eliminated, and computations performed with well-conditioned lower-order matrices.
(e) The design can be extended for the development of corresponding four-stage observers (Sinha 2007; Chen 2012) and filters, as well as for observer- and filter-driven linear-quadratic optimal controllers (it can be also extended to linear stochastic systems) including their four-time scale counterparts.
(f) The design is independent for each local subsystem so that it provides flexibility for the development of partial full-state feedback and partial output feedback controllers including linear-quadratic optimal controllers.
(g) Robustness and reliability can be easily facilitated by using multistage designs, as well as the feedback control loop security can be improved, which appears to be a very important feature these days, especially for cyber physical systems and computer and communication networks.

As an important application of the proposed methodology, we specialize the technique developed in Sect. 6.2 to large-scale linear control systems with slow and fast modes (multi-time scale feedback systems, singularly perturbed control systems; Kokotovic et al. 1999; Naidu and Calise 2001; Dimitriev and Kurina 2006, and Kuehn 2015) for which the proposed type of designs seems to be very well suited. Many systems in mechanical and aerospace engineering possess this structure (Hsiao et al. 2001; Naidu and Calise 2001; Chen et al. 2002; Shapira and Ben-Asher 2004; Demetriou and Kazantzis 2005; Wang and Ghorbel 2006; Amjadifard et al. 2011; Kuehn 2015) due to the presence of small and large time constants, small masses, small moments of inertia, and small stiffness coefficients, which cause clustering of the system eigenvalues into several disjoint groups and particularly into four groups producing a four-time scale linear dynamic system.

6.2 Four-Stage Design of Continuous-Time Feedback Controllers

The efficient methods for the two- and three-stage continuous-time linear feedback controller designs have been presented and demonstrated in Chaps. 2 and 4. In this section, we will extend the continuous-time results of Chap. 4 to the four-stage feedback control design. In the follow-up of the chapter, we will indicate how the newly obtained results can be efficiently used for four-stage feedback control of four-time scale linear control systems (singularly perturbed linear control systems).

Consider a linear continuous-time, time-invariant, dynamic system represented in its partitioned form by

$$\frac{dx(t)}{dt}=\begin{bmatrix}\frac{dx_{\mathrm{I}}(t)}{dt}\\ \frac{dx_{\mathrm{II}}(t)}{dt}\\ \frac{dx_{\mathrm{III}}(t)}{dt}\\ \frac{dx_{\mathrm{IV}}(t)}{dt}\end{bmatrix}=\begin{bmatrix}A_{11} & A_{12} & A_{13} & A_{14}\\ A_{21} & A_{22} & A_{23} & A_{24}\\ A_{31} & A_{32} & A_{33} & A_{34}\\ A_{41} & A_{42} & A_{43} & A_{44}\end{bmatrix}\begin{bmatrix}x_{\mathrm{I}}(t)\\ x_{\mathrm{II}}(t)\\ x_{\mathrm{III}}(t)\\ x_{\mathrm{IV}}(t)\end{bmatrix}+\begin{bmatrix}B_{11}\\ B_{22}\\ B_{33}\\ B_{44}\end{bmatrix}u(t)=Ax(t)+Bu(t)$$

$$y(t)=\begin{bmatrix}C_{11} & C_{22} & C_{33} & C_{44}\end{bmatrix}\begin{bmatrix}x_{\mathrm{I}}(t)\\ x_{\mathrm{II}}(t)\\ x_{\mathrm{III}}(t)\\ x_{\mathrm{IV}}(t)\end{bmatrix}=Cx(t) \tag{6.1}$$

where $x(t) \in \mathrm{R}^n$, $x_{\mathrm{I}}(t)\in\mathrm{R}^{n_1}$, $x_{\mathrm{II}}(t)\in\mathrm{R}^{n_2}$, $x_{\mathrm{III}}(t)\in\mathrm{R}^{n_3}$, $x_{\mathrm{IV}}(t)\in\mathrm{R}^{n_4}$, and $n = n_1 + n_2 + n_3 + n_4$ are system state space variables, $u(t) \in \mathrm{R}^m$ is the system control input vector, $y(t) \in \mathrm{R}^p$ is the vector of system measurements, and A_{ij}, B_{ii}, C_{ii}, and $i, j = 1, 2, 3, 4$ are constant matrices of appropriate dimensions. Matrices A_{11}, A_{22}, A_{33}, and A_{44} define subsystems of dimensions n_1, n_2, n_3, and n_4, respectively, corresponding to the state variables $x_{\mathrm{I}}(t)$, $x_{\mathrm{II}}(t)$, $x_{\mathrm{III}}(t)$, and $x_{\mathrm{IV}}(t)$. Matrices A_{ij}, $i, j = 1, 2, 3, 4,\ i \neq j$ define couplings between the subsystems.

The steps of the four-stage feedback controller design are presented in the follow-up of this section. In the introductory stage, we map the system into appropriate coordinates via several changes of variables in order to achieve the system upper block triangular form.

First Change of Variables

We start with the following change of variables applied to the original system defined in (6.1)

$$\eta_4(t) = L_1 x_{\mathrm{I}}(t) + L_2 x_{\mathrm{II}}(t) + L_3 x_{\mathrm{III}}(t) + x_{\mathrm{IV}}(t) \tag{6.2}$$

which produces the dynamic equation for $\eta_4(t)$ as

$$\begin{aligned}\frac{d\eta_4(t)}{dt} &= (A_{44} + L_1 A_{14} + L_2 A_{24} + L_3 A_{34})\eta_4(t) + (B_{44} + L_1 B_{11} + L_2 B_{22} + L_3 B_{33})u(t) \\ &\quad + f_{41}(L_1, L_2, L_3)x_{\mathrm{I}}(t) + f_{42}(L_1, L_2, L_3)x_{\mathrm{II}}(t) + f_{43}(L_1, L_2, L_3)x_{\mathrm{III}}(t) \\ &= A_4 \eta_4(t) + B_4 u(t) \\ A_4 &= A_{44} + L_1 A_{14} + L_2 A_{24} + L_3 A_{34}, \quad B_4 = B_{44} + L_1 B_{11} + L_2 B_{22} + L_3 B_{33}\end{aligned} \tag{6.3}$$

Elimination of the coupling terms in (6.3) is achieved by assuming that the following systems of algebraic equations have real solutions

$$\begin{aligned} f_{41}(L_1, L_2, L_3) &= L_1 A_{11} + L_2 A_{21} + L_3 A_{31} + A_{41} - (L_1 A_{14} + L_2 A_{24} + L_3 A_{34} + A_{44})L_1 = 0 \\ f_{42}(L_1, L_2, L_3) &= L_1 A_{12} + L_2 A_{22} + L_3 A_{32} + A_{42} - (L_1 A_{14} + L_2 A_{24} + L_3 A_{34} + A_{44})L_2 = 0 \\ f_{43}(L_1, L_2, L_3) &= L_1 A_{13} + L_2 A_{23} + L_3 A_{33} + A_{43} - (L_1 A_{14} + L_2 A_{24} + L_3 A_{34} + A_{44})L_3 = 0 \end{aligned} \tag{6.4}$$

Solving coupled nonlinear algebraic equations (6.4), in general, is not an easy task. However, we will indicate in the next section that for the four-time scale linear systems, the corresponding algebraic equations have much simpler forms so that potentially they can be efficiently solved numerically using either the fixed-point iterations or the Newton algorithm.

Eliminating state variable $x_{\mathrm{IV}}(t)$ from (6.1) using (6.2), the differential equations for $x_{\mathrm{I}}(t)$, $x_{\mathrm{II}}(t)$, and $x_{\mathrm{III}}(t)$ become

$$\frac{dx_{\mathrm{I}}(t)}{dt} = (A_{11} - A_{14}L_1)x_{\mathrm{I}}(t) + (A_{12} - A_{14}L_2)x_{\mathrm{II}}(t) + (A_{13} - A_{14}L_3)x_{\mathrm{III}}(t) + A_{14}\eta_4(t) + B_{11}u(t) \tag{6.5a}$$

$$\frac{dx_{\mathrm{II}}(t)}{dt} = (A_{21} - A_{24}L_1)x_{\mathrm{I}}(t) + (A_{22} - A_{24}L_2)x_{\mathrm{II}}(t) + (A_{23} - A_{24}L_3)x_{\mathrm{III}}(t) + A_{24}\eta_4(t) + B_{22}u(t) \tag{6.5b}$$

$$\frac{dx_{\mathrm{III}}(t)}{dt} = (A_{31} - A_{34}L_1)x_{\mathrm{I}}(t) + (A_{32} - A_{34}L_2)x_{\mathrm{II}}(t) + (A_{33} - A_{34}L_3)x_{\mathrm{III}}(t) + A_{34}\eta_4(t) + B_{33}u(t) \tag{6.5c}$$

Since the $\eta_4(t)$-subsystem (6.3) is isolated, one can design a feedback controller for it using $u(t) = -G_4\eta_4(t) + v(t)$, with $v(t)$ representing an input signal that can be used to control the other subsystems. This strategy was used in the two-stage feedback controller design in Chap. 2. However, in this chapter, due to a more complex four-stage design, we will first achieve an upper block triangular structure for the overall system and then show how to design independently local feedback controllers. The same strategy was also used in Chap. 4, where we presented the three-stage feedback controller design. Before we proceed, we first introduce simplifying notations and present the subsystems defined in (6.5) in the following forms

$$\begin{gathered}\frac{dx_{\mathrm{I}}(t)}{dt} = a_{11}x_{\mathrm{I}}(t) + a_{12}x_{\mathrm{II}}(t) + a_{13}x_{\mathrm{III}}(t) + a_{14}\eta_4(t) + b_{11}u(t) \\ a_{11} = A_{11} - A_{14}L_1, \quad a_{12} = A_{12} - A_{14}L_2, \quad a_{13} = A_{13} - A_{14}L_3, \quad a_{14} = A_{14} \\ b_{11} = B_{11}\end{gathered} \tag{6.6a}$$

$$\begin{gathered}\frac{dx_{\mathrm{II}}(t)}{dt} = a_{21}x_{\mathrm{I}}(t) + a_{22}x_{\mathrm{II}}(t) + a_{23}x_{\mathrm{III}}(t) + a_{24}\eta_4(t) + b_{22}u(t) \\ a_{21} = A_{21} - A_{24}L_1, \quad a_{22} = A_{22} - A_{24}L_2, \quad a_{23} = A_{23} - A_{24}L_3, \quad a_{24} = A_{24} \\ b_{22} = B_{22}\end{gathered} \tag{6.6b}$$

$$\begin{gathered}\frac{dx_{\mathrm{III}}(t)}{dt} = a_{31}x_{\mathrm{I}}(t) + a_{32}x_{\mathrm{II}}(t) + a_{33}x_{\mathrm{III}}(t) + a_{34}\eta_4(t) + b_{33}u(t) \\ a_{31} = A_{31} - A_{34}L_1, \quad a_{32} = A_{32} - A_{34}L_2, \quad a_{33} = A_{33} - A_{34}L_3, \quad a_{34} = A_{34} \\ b_{33} = B_{33}\end{gathered} \tag{6.6c}$$

Second Change of Variables

Now, we introduce the second change of variables to remove dynamics of the other subsystems from the newly defined third subsystem

$$\eta_3(t) = L_4x_{\mathrm{I}}(t) + L_5x_{\mathrm{II}}(t) + x_{\mathrm{III}}(t) \tag{6.7}$$

This change of variables leads to the following dynamic system for $\eta_3(t)$

$$\begin{aligned}\frac{d\eta_3(t)}{dt} &= (a_{33} + L_4a_{13} + L_5a_{23})\eta_3(t) + (a_{34} + L_4a_{14} + L_5a_{24})\eta_4(t) \\ &\quad + (b_{33} + L_4b_{11} + L_5b_{22})u(t) + f_{31}(L_4, L_5)x_{\mathrm{I}}(t) + f_{32}(L_4, L_5)x_{\mathrm{II}}(t) \\ &= (a_{33} + L_4a_{13} + L_5a_{23})\eta_3(t) + (a_{34} + L_4a_{14} + L_5a_{24})\eta_4(t) \\ &\quad + (b_{33} + L_4b_{11} + L_5b_{22})u(t)\end{aligned} \tag{6.8}$$

In (6.8), the following result was used $f_{31}(L_4, L_5) = 0$ and $f_{32}(L_4, L_5) = 0$. These two algebraic equations are given by

$$\begin{aligned}f_{31}(L_4, L_5) &= (a_{31} + L_4a_{11} + L_5a_{21}) - (a_{33} + L_4a_{13} + L_5a_{23})L_4 = 0 \\ f_{32}(L_4, L_5) &= (a_{32} + L_4a_{12} + L_5a_{22}) - (a_{33} + L_4a_{13} + L_5a_{23})L_5 = 0\end{aligned} \tag{6.9}$$

In general, it is not easy to solve the nonlinear algebraic equations defined in (6.9). How to efficiently solve these algebraic equations will be considered in the second part of this chapter for a special case of four-time scale linear dynamic systems.

After the change of variables defined in (6.7), the first and second subsystems in the new coordinates become

$$\begin{aligned}\frac{dx_{\mathrm{I}}(t)}{dt} &= \alpha_{11}x_{\mathrm{I}}(t) + \alpha_{12}x_{\mathrm{II}}(t) + \alpha_{13}\eta_3(t) + \alpha_{14}\eta_4(t) + \beta_{11}u(t) \\ \frac{dx_{\mathrm{II}}(t)}{dt} &= \alpha_{21}x_{\mathrm{I}}(t) + \alpha_{22}x_{\mathrm{II}}(t) + \alpha_{23}\eta_3(t) + \alpha_{24}\eta_4(t) + \beta_{22}u(t)\end{aligned} \tag{6.10a}$$

where the newly introduced matrices α_{ij}, $i = 1, 2, j = 1, 2, 3, 4$, and β_{ii} are defined by the following expressions

$$\begin{aligned}&\alpha_{11} = a_{11} - a_{13}L_4, \quad \alpha_{12} = a_{12} - a_{13}L_5, \quad \alpha_{13} = a_{13}, \quad \alpha_{14} = a_{14}, \quad \beta_{11} = b_{11} \\ &\alpha_{21} = a_{21} - a_{23}L_4, \quad \alpha_{22} = a_{22} - a_{23}L_5, \quad \alpha_{23} = a_{23}, \quad \alpha_{24} = a_{24}, \quad \beta_{22} = b_{22}\end{aligned} \tag{6.10b}$$

Third Change of Variables

Another change of variables is needed to eliminate $x_{\mathrm{II}}(t)$ from (6.10a). This transformation is given by

$$\eta_2(t) = L_6x_{\mathrm{I}}(t) + x_{\mathrm{II}}(t) \tag{6.11}$$

In the new coordinates, we have the following subsystem for $\eta_2(t)$

$$\begin{aligned}\frac{d\eta_2(t)}{dt} = &(\alpha_{22} + L_6\alpha_{12})\eta_2(t) + (\alpha_{23} + L_6\alpha_{13})\eta_3(t) + (\alpha_{24} + L_6\alpha_{14})\eta_4(t) \\ &+ (\beta_{22} + L_6\beta_{11})u(t) + f_{22}(L_6)x_{\mathrm{I}}(t)\end{aligned} \tag{6.12}$$

where

$$f_{21}(L_6) = \alpha_{21} + L_6\alpha_{11} - (\alpha_{22} + L_6\alpha_{12})L_6 = 0 \tag{6.13}$$

It is expected that the equation for L_6 can be solved efficiently so that the state variable $x_{\mathrm{I}}(t)$ is eliminated from the differential equation (6.12).

Solving (6.11) for $x_{\mathrm{II}}(t)$, that is, $x_{\mathrm{II}}(t) = \eta_2(t) - L_6x_{\mathrm{I}}(t)$ and eliminating $x_{\mathrm{II}}(t)$ from the first equation in (6.10a), we obtain

$$\frac{dx_{\mathrm{I}}(t)}{dt} = (\alpha_{11} - \alpha_{12}L_6)x_{\mathrm{I}}(t) + \alpha_{12}\eta_2(t) + \alpha_{13}\eta_3(t) + \alpha_{14}\eta_4(t) + \beta_{11}u(t) \tag{6.14}$$

Linear dynamic systems defined by differential equations (6.3), (6.8), (6.12), and (6.14) represent the sought upper block triangular form whose state space form is given by

$$\begin{bmatrix} \frac{dx_{\mathrm{I}}(t)}{dt} \\ \frac{d\eta_2(t)}{dt} \\ \frac{d\eta_3(t)}{dt} \\ \frac{d\eta_4(t)}{dt} \end{bmatrix} = \begin{bmatrix} \alpha_{11} - \alpha_{12}L_6 & \alpha_{12} & \alpha_{13} & \alpha_{14} \\ 0 & \alpha_{22} + L_6\alpha_{12} & \alpha_{23} + L_6\alpha_{13} & \alpha_{24} + L_6\alpha_{14} \\ 0 & 0 & \alpha_{33} & \alpha_{34} \\ 0 & 0 & 0 & A_4 \end{bmatrix} \begin{bmatrix} x_{\mathrm{I}}(t) \\ \eta_2(t) \\ \eta_3(t) \\ \eta_4(t) \end{bmatrix}$$
$$+ \begin{bmatrix} \beta_{11} \\ \beta_{22} + L_6\beta_{11} \\ \beta_{33} \\ B_4 \end{bmatrix} u(t) = \bar{A} \begin{bmatrix} x_{\mathrm{I}}(t) \\ \eta_2(t) \\ \eta_3(t) \\ \eta_4(t) \end{bmatrix} + \bar{B}u(t) \tag{6.15}$$

The changes of the state space variables used in this process given in (6.2), (6.7), and (6.11) relate the original and new state variables via the following similarity transformation

$$\begin{bmatrix} x_{\mathrm{I}}(t) \\ \eta_2(t) \\ \eta_3(t) \\ \eta_4(t) \end{bmatrix} = \begin{bmatrix} I & 0 & 0 & 0 \\ L_6 & I & 0 & 0 \\ L_4 & L_5 & I & 0 \\ L_1 & L_2 & L_3 & I \end{bmatrix} \begin{bmatrix} x_{\mathrm{I}}(t) \\ x_{\mathrm{II}}(t) \\ x_{\mathrm{III}}(t) \\ x_{\mathrm{IV}}(t) \end{bmatrix} = T_1 \begin{bmatrix} x_{\mathrm{I}}(t) \\ x_{\mathrm{II}}(t) \\ x_{\mathrm{III}}(t) \\ x_{\mathrm{IV}}(t) \end{bmatrix} \tag{6.16}$$

The inverse transformation of (6.16) is given by

$$\begin{bmatrix} x_{\mathrm{I}}(t) \\ x_{\mathrm{II}}(t) \\ x_{\mathrm{III}}(t) \\ x_{\mathrm{IV}}(t) \end{bmatrix} = \begin{bmatrix} I & 0 & 0 & 0 \\ -L_6 & I & 0 & 0 \\ L_5L_6 - L_4 & -L_5 & I & 0 \\ L_3L_4 + L_2L_6 - L_1 - L_3L_5L_6 & -L_3L_5 - L_2 & -L_3 & I \end{bmatrix} \begin{bmatrix} x_{\mathrm{I}}(t) \\ \eta_2(t) \\ \eta_3(t) \\ \eta_4(t) \end{bmatrix}$$
$$= T_1^{-1} \begin{bmatrix} x_{\mathrm{I}}(t) \\ \eta_2(t) \\ \eta_3(t) \\ \eta_4(t) \end{bmatrix} \tag{6.17}$$

Having transformed the original system into a lower block triangular form, we can start the design of feedback controllers. We will indicate intermediate steps needed to make these designs independent. Before we proceed, we simplify notation and make it more uniform for the derived system (6.15) as follows:

$$\begin{bmatrix} \frac{dx_1(t)}{dt} \\ \frac{d\eta_2(t)}{dt} \\ \frac{d\eta_3(t)}{dt} \\ \frac{d\eta_4(t)}{dt} \end{bmatrix} = \bar{A} \begin{bmatrix} x_1(t) \\ \eta_2(t) \\ \eta_3(t) \\ \eta_4(t) \end{bmatrix} + \bar{B}u(t) = \begin{bmatrix} \bar{A}_{11} & \bar{A}_{12} & \bar{A}_{13} & \bar{A}_{14} \\ 0 & \bar{A}_{22} & \bar{A}_{23} & \bar{A}_{24} \\ 0 & 0 & \bar{A}_{33} & \bar{A}_{34} \\ 0 & 0 & 0 & \bar{A}_{44} \end{bmatrix} \begin{bmatrix} x_1(t) \\ \eta_2(t) \\ \eta_3(t) \\ \eta_4(t) \end{bmatrix} + \begin{bmatrix} \bar{B}_{11} \\ \bar{B}_{22} \\ \bar{B}_{33} \\ \bar{B}_{44} \end{bmatrix} u(t)$$

$$\begin{aligned} &\bar{A}_{11} = \alpha_{11} - \alpha_{12}L_6, \quad \bar{A}_{12} = \alpha_{12}, \quad \bar{A}_{13} = \alpha_{13}, \quad \bar{A}_{14} = \alpha_{14} \\ &\bar{A}_{22} = \alpha_{22} + L_6\alpha_{12}, \quad \bar{A}_{23} = \alpha_{23} + L_6\alpha_{13}, \quad \bar{A}_{24} = \alpha_{24} + L_6\alpha_{14} \\ &\bar{A}_{33} = \alpha_{33}, \quad \bar{A}_{34} = \alpha_{34}, \quad \bar{A}_{44} = A_4 \\ &\bar{B}_{11} = \beta_{11}, \quad \bar{B}_{22} = \beta_{22} + L_6\beta_{11}, \quad \bar{B}_{33} = \beta_{33}, \quad \bar{B}_{44} = B_4 \end{aligned} \tag{6.18}$$

Now we start the four-stage feedback controller design procedure, in which we isolate every subsystem and identify the control input that independently of other subsystems solely controls the isolated subsystem.

Stage 1

Applying feedback control for the fourth subsystem as

$$u(t) = -G_4\eta_4(t) + v(t) \tag{6.19}$$

the overall system (6.18) becomes

$$\begin{aligned} \begin{bmatrix} \frac{dx_1(t)}{dt} \\ \frac{d\eta_2(t)}{dt} \\ \frac{d\eta_3(t)}{dt} \\ \frac{d\eta_4(t)}{dt} \end{bmatrix} &= \bar{A} \begin{bmatrix} x_1(t) \\ \eta_2(t) \\ \eta_3(t) \\ \eta_4(t) \end{bmatrix} + \bar{B}u(t) \\ &= \begin{bmatrix} \bar{A}_{11} & \bar{A}_{12} & \bar{A}_{13} & \bar{A}_{14} - \bar{B}_{11}G_4 \\ 0 & \bar{A}_{22} & \bar{A}_{23} & \bar{A}_{24} - \bar{B}_{22}G_4 \\ 0 & 0 & \bar{A}_{33} & \bar{A}_{34} - \bar{B}_{33}G_4 \\ 0 & 0 & 0 & \bar{A}_{44} - \bar{B}_{44}G_4 \end{bmatrix} \begin{bmatrix} x_1(t) \\ \eta_2(t) \\ \eta_3(t) \\ \eta_4(t) \end{bmatrix} + \begin{bmatrix} \bar{B}_{11} \\ \bar{B}_{22} \\ \bar{B}_{33} \\ \bar{B}_{44} \end{bmatrix} v(t) \end{aligned} \tag{6.20}$$

Stage 2

To isolate the third subsystem, we map the system into the new coordinates and apply the following change of variables as

$$\xi_3(t) = \eta_3(t) - P_3\eta_4(t) \tag{6.21}$$

which produces

$$\begin{bmatrix} \frac{dx_1(t)}{dt} \\ \frac{d\eta_2(t)}{dt} \\ \frac{d\xi_3(t)}{dt} \\ \frac{d\eta_4(t)}{dt} \end{bmatrix} = \begin{bmatrix} \bar{A}_{11} & \bar{A}_{12} & \bar{A}_{13} & \bar{A}_{14} - \bar{B}_{11}G_4 + \bar{A}_{13}P_3 \\ 0 & \bar{A}_{22} & \bar{A}_{23} & \bar{A}_{24} - \bar{B}_{22}G_4 + \bar{A}_{23}P_3 \\ 0 & 0 & \bar{A}_{33} & 0 \\ 0 & 0 & 0 & \bar{A}_{44} - \bar{B}_{44}G_4 \end{bmatrix} \begin{bmatrix} x_1(t) \\ \eta_2(t) \\ \xi_3(t) \\ \eta_4(t) \end{bmatrix} + \begin{bmatrix} \bar{B}_{11} \\ \bar{B}_{22} \\ \bar{B}_{33} - P_3\bar{B}_{44} \\ \bar{B}_{44} \end{bmatrix} v(t) \tag{6.22}$$

In (6.22), we have used the fact that matrix P_3 satisfies the following algebraic equation

$$\bar{A}_{33}P_3 - P_3(\bar{A}_{44} - \bar{B}_{44}G_4) + \bar{A}_{34} - \bar{B}_{33}G_4 = 0 \tag{6.23}$$

Note that (6.23) is the Sylvester algebraic equation, and its unique solution exists under the following assumption, Chen (2012).

Assumption 6.1 Matrices $\bar{A}_{33}$ and $\bar{A}_{44} - \bar{B}_{44}G_4$ have no eigenvalues in common.

This assumption is easily satisfied since $\bar{A}_{44} - \bar{B}_{44}G_4$ is the feedback matrix of the fourth subsystem, and the Sylvester algebraic equation (6.23) can be solved directly.

Since the third subsystem is isolated and not connected to the remaining subsystems, in this stage, we apply linear feedback control using the state variables from the third subsystem only as

$$v(t) = -G_3\xi_3(t) + w(t) \tag{6.24}$$

which produces the following system

$$\begin{bmatrix} \frac{dx_{\mathrm{I}}(t)}{dt} \\ \frac{d\eta_2(t)}{dt} \\ \frac{d\xi_3(t)}{dt} \\ \frac{d\eta_4(t)}{dt} \end{bmatrix} = \begin{bmatrix} \bar{A}_{11} & \bar{A}_{12} & \bar{A}_{13} - \bar{B}_{11}G_3 & \bar{A}_{14} - \bar{B}_{11}G_4 + \bar{A}_{13}P_3 \\ 0 & \bar{A}_{22} & \bar{A}_{23} - \bar{B}_{22}G_3 & \bar{A}_{24} - \bar{B}_{22}G_4 + \bar{A}_{23}P_3 \\ 0 & 0 & \bar{A}_{33} - \left(\bar{B}_{33} - P_3\bar{B}_{44}\right)G_3 & 0 \\ 0 & 0 & -\bar{B}_{44}G_3 & \bar{A}_{44} - \bar{B}_{44}G_4 \end{bmatrix} \begin{bmatrix} x_{\mathrm{I}}(t) \\ \eta_2(t) \\ \xi_3(t) \\ \eta_4(t) \end{bmatrix} + \begin{bmatrix} \bar{B}_{11} \\ \bar{B}_{22} \\ \bar{B}_{33} - P_3\bar{B}_{44} \\ \bar{B}_{44} \end{bmatrix} v(t) \tag{6.25}$$

Stage 3
Now we need to isolate the second subsystem and remove its coupling from the remaining subsystems. To achieve that goal, we apply another change of state space variables as

$$\xi_2(t) = \eta_2(t) - P_{23}\xi_3(t) - P_{24}\eta_4(t) \tag{6.26}$$

where P_{23} and P_{24} satisfy the following system of two linear matrix algebraic equations

$$\bar{A}_{22}P_{23} - P_{23}\left(\bar{A}_{33} - (\bar{B}_{33} - P_3\bar{B}_{44})G_3\right) + P_{24}\bar{B}_{44}G_3 + \bar{A}_{23} - \bar{B}_{22}G_3 = 0 \tag{6.27}$$

$$\bar{A}_{22}P_{24} - P_{24}(\bar{A}_{44} - \bar{B}_{44}G_4) + \bar{A}_{24} - \bar{B}_{22}G_4 + \bar{A}_{23}P_3 = 0 \tag{6.28}$$

Equations (6.27) and (6.28) are Sylvester algebraic equations, and their unique solution exists under the assumptions that the eigenvalues of the corresponding coefficient matrices do not sum up to zero (Chen 2012). Hence, the following assumptions are needed:

Assumption 6.2a Matrices $\bar{A}_{22}$ and $\bar{A}_{33} - \left(\bar{B}_{33} - P_3\bar{B}_{44}\right)G_3$ have no eigenvalues in common.

Assumption 6.2b The matrix $\bar{A}_{22}$ and matrix $(\bar{A}_{44} - \bar{B}_{44}G_4)$ have no eigenvalues in common.

The change of variables defined in (6.26) modifies the linear dynamic system (6.25) into

$$
\begin{bmatrix} \frac{dx_{\mathrm{I}}(t)}{dt} \\ \frac{d\xi_2(t)}{dt} \\ \frac{d\xi_3(t)}{dt} \\ \frac{d\eta_4(t)}{dt} \end{bmatrix} = \begin{bmatrix} \bar{A}_{11} & \bar{A}_{12} & \bar{A}_{13} - \bar{B}_{11}G_3 + \bar{A}_{12}P_{23} & S_{14} \\ 0 & \bar{A}_{22} & 0 & 0 \\ 0 & 0 & \bar{A}_{33} - (\bar{B}_{33} - P_3\bar{B}_{44})G_3 & 0 \\ 0 & 0 & -\bar{B}_{44}G_3 & \bar{A}_{44} - \bar{B}_{44}G_4 \end{bmatrix} \begin{bmatrix} x_{\mathrm{I}}(t) \\ \xi_2(t) \\ \xi_3(t) \\ \eta_4(t) \end{bmatrix}
$$

$$
+ \begin{bmatrix} \bar{B}_{11} \\ \bar{B}_{22} - P_{23}(\bar{B}_{33} - P_3\bar{B}_{44}) - P_{24}\bar{B}_{44} \\ \bar{B}_{33} - P_3\bar{B}_{44} \\ \bar{B}_{44} \end{bmatrix} w(t) \tag{6.29a}
$$

where

$$
S_{14} = \bar{A}_{14} - \bar{B}_{11}G_4 + \bar{A}_{13}P_3 + \bar{A}_{12}P_{24} \tag{6.29b}
$$

Since the second subsystem is isolated, the local state feedback input can be used to independently control this subsystem

$$
w(t) = -G_2\xi_2(t) + f(t) \tag{6.30}
$$

which leads to the following overall system, obtained by closing the feedback loop for the second subsystem and appropriately controlling the second subsystem using partial state feedback generated by using only the state space variables of the second subsystem

$$
\begin{bmatrix} \frac{dx_{\mathrm{I}}(t)}{dt} \\ \frac{d\xi_2(t)}{dt} \\ \frac{d\xi_3(t)}{dt} \\ \frac{d\eta_4(t)}{dt} \end{bmatrix} = \begin{bmatrix} \bar{A}_{11} & \bar{A}_{12} - \bar{B}_{11}G_2 & \bar{A}_{13} - \bar{B}_{11}G_3 + \bar{A}_{12}P_{23} & S_{14} \\ 0 & \bar{A}_{22} - \bar{B}_{2\xi}G_2 & 0 & 0 \\ 0 & -\bar{B}_{3\xi}G_2 & \bar{A}_{33} - (\bar{B}_{33} - P_3\bar{B}_{44})G_3 & 0 \\ 0 & -\bar{B}_{44}G_2 & -\bar{B}_{44}G_3 & \bar{A}_{44} - \bar{B}_{44}G_4 \end{bmatrix} \begin{bmatrix} x_{\mathrm{I}}(t) \\ \xi_2(t) \\ \xi_3(t) \\ \eta_4(t) \end{bmatrix}
$$

$$
+ \begin{bmatrix} \bar{B}_{11} \\ \bar{B}_{2\xi} \\ \bar{B}_{3\xi} \\ \bar{B}_{44} \end{bmatrix} f(t) \tag{6.31a}
$$

where

$$\bar{B}_{2\xi} = \bar{B}_{22} - P_{23}(\bar{B}_{33} - P_3\bar{B}_{44}) - P_{24}\bar{B}_{44}, \quad \bar{B}_{3\xi} = \bar{B}_{33} - P_3\bar{B}_{44} \tag{6.31b}$$

Before proceeding to Stage 4 (where we will show how to make (6.31a) an upper block triangular system, which will make the first subsystem dynamically isolated from the remaining subsystems so its local state feedback controller can be independently designed), we first simplify notation for the system derived in (6.31) and represent it by

$$\begin{bmatrix} \frac{dx_1(t)}{dt} \\ \frac{d\xi_2(t)}{dt} \\ \frac{d\xi_3(t)}{dt} \\ \frac{d\eta_4(t)}{dt} \end{bmatrix} = \begin{bmatrix} S_{11} & S_{12} & S_{13} & S_{14} \\ 0 & S_{22} & 0 & 0 \\ 0 & S_{32} & S_{33} & 0 \\ 0 & S_{42} & S_{43} & S_{44} \end{bmatrix} \begin{bmatrix} x_1(t) \\ \xi_2(t) \\ \xi_3(t) \\ \eta_4(t) \end{bmatrix} + \begin{bmatrix} Q_1 \\ Q_2 \\ Q_3 \\ Q_4 \end{bmatrix} f(t) \tag{6.32a}$$

where

$$\begin{aligned} &S_{11} = \bar{A}_{11}, \quad S_{12} = \bar{A}_{12} - \bar{B}_{11}G_2, \quad S_{13} = \bar{A}_{13} - \bar{B}_{11}G_3 + \bar{A}_{12}P_{23} \\ &S_{22} = \bar{A}_{22} - \bar{B}_{2\xi}G_2 \\ &S_{32} = -\bar{B}_{3\xi}G_2, \quad S_{33} = \bar{A}_{33} - (\bar{B}_{33} - P_3\bar{B}_{44})G_3 \\ &S_{42} = -\bar{B}_{44}G_2, \quad S_{43} = -\bar{B}_{44}G_3, \quad S_{44} = \bar{A}_{44} - \bar{B}_{44}G_4 \\ &Q_1 = \bar{B}_{11}, \quad Q_2 = \bar{B}_{22} - P_{23}(\bar{B}_{33} - P_3\bar{B}_{44}) - P_{24}\bar{B}_{44}, \\ &Q_3 = \bar{B}_{33} - P_3\bar{B}_{44}, \quad Q_4 = \bar{B}_{44} \end{aligned} \tag{6.32b}$$

with S_{14} defined in (6.29b).

Stage 4
Introduce the following change of the state space variables

$$\xi_1(t) = x_1(t) - P_{12}\xi_2(t) - P_{13}\xi_3(t) - P_{14}\eta_4(t) \tag{6.33}$$

where matrices P_{12},P_{13},P_{14} satisfy the following system of linear algebraic equations

$$\begin{aligned} &S_{11}P_{12} - P_{12}S_{22} - P_{13}S_{32} - P_{14}S_{42} + S_{12} = 0 \\ &S_{11}P_{13} - P_{13}S_{33} - P_{14}S_{43} + S_{13} = 0 \\ &S_{11}P_{14} - P_{14}S_{44} + S_{14} = 0 \end{aligned} \tag{6.34}$$

The algebraic equation for P_{14} is the Sylvester algebraic equation whose unique solution exists under the following assumption:

Assumption 6.3a Matrices S_{11} and S_{44} have no eigenvalues in common.

Having obtained the solution for P_{14}, the algebraic equation for P_{13} is another Sylvester algebraic equation whose unique solution exists under the assumption.

Assumption 6.3b Matrices S_{11} and S_{33} have no eigenvalues in common.

Having obtained the solutions for both P_{14} and P_{13}, the algebraic equation for P_{12} is the algebraic Sylvester equation that requires the following assumption.

Assumption 6.3c Matrices S_{11} and S_{22} have no eigenvalues in common.

Applying (6.33) to the system (6.32a), in the new coordinates, we obtain the following system:

$$\begin{bmatrix} \frac{d\xi_1(t)}{dt} \\ \frac{d\xi_2(t)}{dt} \\ \frac{d\xi_3(t)}{dt} \\ \frac{d\eta_4(t)}{dt} \end{bmatrix} = \begin{bmatrix} S_{11} & 0 & 0 & 0 \\ 0 & S_{22} & 0 & 0 \\ 0 & S_{32} & S_{33} & 0 \\ 0 & S_{42} & S_{43} & S_{44} \end{bmatrix} \begin{bmatrix} \xi_1(t) \\ \xi_2(t) \\ \xi_3(t) \\ \eta_4(t) \end{bmatrix} + \begin{bmatrix} Q_1 - P_{12}Q_2 - P_{13}Q_3 - P_{14}Q_4 \\ Q_2 \\ Q_3 \\ Q_4 \end{bmatrix} f(t) \tag{6.35}$$

The obtained system matrix is block upper triangular so its closed-loop eigenvalues are the union of the closed-loop eigenvalues of the individual subsystems. The first subsystem is isolated, and it can be independently controlled via its own state feedback

$$f(t) = -G_1\xi_1(t) \tag{6.36}$$

which produces the following closed-loop system

$$\begin{bmatrix} \frac{d\xi_1(t)}{dt} \\ \frac{d\xi_2(t)}{dt} \\ \frac{d\xi_3(t)}{dt} \\ \frac{d\eta_4(t)}{dt} \end{bmatrix} = \begin{bmatrix} S_{11} - (Q_1 - P_{12}Q_2 - P_{13}Q_3 - P_{14}Q_4)G_1 & 0 & 0 & 0 \\ -Q_2G_1 & S_{22} & 0 & 0 \\ -Q_3G_1 & S_{32} & S_{33} & 0 \\ -Q_4G_1 & S_{42} & S_{43} & S_{44} \end{bmatrix} \begin{bmatrix} \xi_1(t) \\ \xi_2(t) \\ \xi_3(t) \\ \eta_4(t) \end{bmatrix} \tag{6.37}$$

We can now relate the new and the original system state variables via a similarity transformation. The state coordinates defined in (6.18) and (6.35) are related via relations (6.21), (6.26), and (6.33). These relationships can be put in a compact matrix form providing a general transformation to the original state variables. First, we use the change of state space variables defined in (6.21) and (6.26) to establish the relation between the systems defined in (6.15) and (6.29). Note that the linear dynamic system (6.15) is related to the original linear dynamic system (6.1) via the similarity transformation derived in (6.16) and (6.17). From (6.21) and (6.26), we have

$$
\begin{aligned}
\begin{bmatrix} x_{\mathrm{I}}(t) \\ \eta_2(t) \\ \xi_3(t) \\ \eta_4(t) \end{bmatrix} &= \begin{bmatrix} I & 0 & 0 & 0 \\ 0 & I & 0 & 0 \\ 0 & 0 & I & -P_3 \\ 0 & 0 & 0 & I \end{bmatrix} \begin{bmatrix} x_{\mathrm{I}}(t) \\ \eta_2(t) \\ \eta_3(t) \\ \eta_4(t) \end{bmatrix} = T_2 \begin{bmatrix} x_{\mathrm{I}}(t) \\ \eta_2(t) \\ \eta_3(t) \\ \eta_4(t) \end{bmatrix} \\
\begin{bmatrix} x_{\mathrm{I}}(t) \\ \eta_2(t) \\ \eta_3(t) \\ \eta_4(t) \end{bmatrix} &= \begin{bmatrix} I & 0 & 0 & 0 \\ 0 & I & 0 & 0 \\ 0 & 0 & I & P_3 \\ 0 & 0 & 0 & I \end{bmatrix} \begin{bmatrix} x_{\mathrm{I}}(t) \\ \eta_2(t) \\ \xi_3(t) \\ \eta_4(t) \end{bmatrix} = T_2^{-1} \begin{bmatrix} x_{\mathrm{I}}(t) \\ \eta_2(t) \\ \xi_3(t) \\ \eta_4(t) \end{bmatrix} \\
\begin{bmatrix} x_{\mathrm{I}}(t) \\ \xi_2(t) \\ \xi_3(t) \\ \eta_4(t) \end{bmatrix} &= \begin{bmatrix} I & 0 & 0 & 0 \\ 0 & I & -P_{23} & -P_{24} \\ 0 & 0 & I & 0 \\ 0 & 0 & 0 & I \end{bmatrix} \begin{bmatrix} x_{\mathrm{I}}(t) \\ \eta_2(t) \\ \xi_3(t) \\ \eta_4(t) \end{bmatrix} = T_3 \begin{bmatrix} x_{\mathrm{I}}(t) \\ \eta_2(t) \\ \xi_3(t) \\ \eta_4(t) \end{bmatrix} \\
\begin{bmatrix} x_{\mathrm{I}}(t) \\ \eta_2(t) \\ \xi_3(t) \\ \eta_4(t) \end{bmatrix} &= \begin{bmatrix} I & 0 & 0 & 0 \\ 0 & I & P_{23} & P_{24} \\ 0 & 0 & I & 0 \\ 0 & 0 & 0 & I \end{bmatrix} \begin{bmatrix} x_{\mathrm{I}}(t) \\ \xi_2(t) \\ \xi_3(t) \\ \eta_4(t) \end{bmatrix} = T_3^{-1} \begin{bmatrix} x_{\mathrm{I}}(t) \\ \xi_2(t) \\ \xi_3(t) \\ \eta_4(t) \end{bmatrix}
\end{aligned} \tag{6.38}
$$

The change of variables defined in (6.33) provides the following relationship

$$
\begin{aligned}
\begin{bmatrix} \xi_1(t) \\ \xi_2(t) \\ \xi_3(t) \\ \eta_4(t) \end{bmatrix} &= \begin{bmatrix} I & -P_{12} & -P_{13} & -P_{14} \\ 0 & I & 0 & 0 \\ 0 & 0 & I & 0 \\ 0 & 0 & 0 & I \end{bmatrix} \begin{bmatrix} x_{\mathrm{I}}(t) \\ \xi_2(t) \\ \xi_3(t) \\ \eta_4(t) \end{bmatrix} = T_4 \begin{bmatrix} x_{\mathrm{I}}(t) \\ \xi_2(t) \\ \xi_3(t) \\ \eta_4(t) \end{bmatrix} \\
\begin{bmatrix} x_{\mathrm{I}}(t) \\ \xi_2(t) \\ \xi_3(t) \\ \eta_4(t) \end{bmatrix} &= \begin{bmatrix} I & P_{12} & P_{13} & P_{14} \\ 0 & I & 0 & 0 \\ 0 & 0 & I & 0 \\ 0 & 0 & 0 & I \end{bmatrix} \begin{bmatrix} \xi_1(t) \\ \xi_2(t) \\ \xi_3(t) \\ \eta_4(t) \end{bmatrix} = T_4^{-1} \begin{bmatrix} \xi_1(t) \\ \xi_2(t) \\ \xi_3(t) \\ \eta_4(t) \end{bmatrix}
\end{aligned} \tag{6.39}
$$

So that the original coordinates (6.1) and the final design coordinates (6.35) are related by

$$
\begin{aligned}
\begin{bmatrix} x_{\mathrm{I}}(t) \\ x_{\mathrm{II}}(t) \\ x_{\mathrm{III}}(t) \\ x_{\mathrm{IV}}(t) \end{bmatrix} &= T_1^{-1} \begin{bmatrix} x_{\mathrm{I}}(t) \\ \eta_2(t) \\ \eta_3(t) \\ \eta_4(t) \end{bmatrix} = T_1^{-1} T_2^{-1} \begin{bmatrix} x_{\mathrm{I}}(t) \\ \eta_2(t) \\ \xi_3(t) \\ \eta_4(t) \end{bmatrix} = T_1^{-1} T_2^{-1} T_3^{-1} \begin{bmatrix} \eta_1(t) \\ \xi_2(t) \\ \xi_3(t) \\ \eta_4(t) \end{bmatrix} \\
&= T_1^{-1} T_2^{-1} T_3^{-1} T_4^{-1} \begin{bmatrix} \xi_1(t) \\ \xi_2(t) \\ \xi_3(t) \\ \eta_4(t) \end{bmatrix} = T \begin{bmatrix} \xi_1(t) \\ \xi_2(t) \\ \xi_3(t) \\ \eta_4(t) \end{bmatrix}
\end{aligned} \tag{6.40}
$$

and the corresponding inverse transformation

$$\begin{bmatrix} \xi_1(t) \\ \xi_2(t) \\ \xi_3(t) \\ \eta_4(t) \end{bmatrix} = T_4T_3T_2T_1 \begin{bmatrix} x_{\mathrm{I}}(t) \\ x_{\mathrm{II}}(t) \\ x_{\mathrm{III}}(t) \\ x_{\mathrm{IV}}(t) \end{bmatrix} = T \begin{bmatrix} x_{\mathrm{I}}(t) \\ x_{\mathrm{II}}(t) \\ x_{\mathrm{III}}(t) \\ x_{\mathrm{IV}}(t) \end{bmatrix} \tag{6.41}$$

The feedback control signal applied to the system in the transformed coordinates is given by

$$\begin{aligned} u(\xi_1(t), \xi_2(t), \xi_3(t), \eta_4(t)) &= -G_1\xi_1(t) - G_2\xi_2(t) - G_3\xi_3(t) - G_4\eta_4(t) \\ &= -[G_1 \quad G_2 \quad G_3 \quad G_4] \begin{bmatrix} \xi_1(t) \\ \xi_2(t) \\ \xi_3(t) \\ \eta_4(t) \end{bmatrix} \\ &= -[G_1 \quad G_2 \quad G_3 \quad G_4]T \begin{bmatrix} x_{\mathrm{I}}(t) \\ x_{\mathrm{II}}(t) \\ x_{\mathrm{III}}(t) \\ x_{\mathrm{IV}}(t) \end{bmatrix} \\ &= -[G_{1eq} \quad G_{2eq} \quad G_{3eq} \quad G_{4eq}] \begin{bmatrix} x_{\mathrm{I}}(t) \\ x_{\mathrm{II}}(t) \\ x_{\mathrm{III}}(t) \\ x_{\mathrm{IV}}(t) \end{bmatrix} \end{aligned} \tag{6.42}$$

with the similarity transformation matrix defined by

$$T = T_4T_3T_2T_1 \tag{6.43}$$

From (6.42), we obtain the feedback full-state control signal in the original coordinates as

$$\begin{aligned} u\left(x_I(t)^I, x_{II}(t)^{II}, x_{III}(t)^{III}, x_{IV}(t)^{IV}\right) &= -G_{1eq}x_I(t)^I - G_{2eq}x_{II}(t)^{II} \\ &\quad - G_{3eq}x_{III}(t)^{III} - G_{4eq}x_{IV}(t)^{IV} \end{aligned} \tag{6.44}$$

6.3 Four-Stage Four-Time Scale Linear Control Systems

Four-time scale systems are an excellent candidate for efficient application of the four-stage linear feedback controllers. For this class of systems, in general, numerical ill-conditioning appears if one attempts to design a linear feedback controller directly using the entire (full-order) system. As indicated in the introductory section, singularly perturbed control systems have numerous applications in all areas of engineering and sciences. In this section, we will indicate how further design simplifications can be achieved by specializing the proposed design from Sect. 6.2

to singularly perturbed continuous-time linear systems. Moreover, the digital implementation of the corresponding controllers will allow different sampling periods to be used for different controllers. Otherwise, without the four-stage design, the whole system digital controller will require a very small sampling period (very large sampling rate).

Several multi-time scales are present in many real physical systems that have components of different nature. For example, advanced heavy water reactor (Shimjith et al. 2011a, b; Munje et al. 2014, 2015a, b) has three time scales. Dynamics of fuel cells evolves in at least three, possibly four, time scales (Prukrushpan et al. 2004a; Zenith and Skogestad 2009). It was shown in Zenith and Skogestad (2009) that a proton-exchange membrane fuel cell (PEMFC) system has three and possibly four subsystems operating in three or four different time scales.

Consider a four-time scale, time-invariant, linear dynamic control system represented by

$$\begin{bmatrix} \dfrac{dx_{\mathrm{I}}(t)}{dt} \\ \varepsilon_1 \dfrac{dx_{\mathrm{II}}(t)}{dt} \\ \varepsilon_2 \dfrac{dx_{\mathrm{III}}(t)}{dt} \\ \varepsilon_3 \dfrac{dx_{\mathrm{IV}}(t)}{dt} \end{bmatrix} = \begin{bmatrix} A_{11} & A_{12} & A_{13} & A_{14} \\ A_{21} & A_{22} & A_{23} & A_{24} \\ A_{31} & A_{32} & A_{33} & A_{34} \\ A_{41} & A_{42} & A_{43} & A_{44} \end{bmatrix} \begin{bmatrix} x_{\mathrm{I}}(t) \\ x_{\mathrm{II}}(t) \\ x_{\mathrm{III}}(t) \\ x_{\mathrm{IV}}(t) \end{bmatrix} + \begin{bmatrix} B_{11} \\ B_{22} \\ B_{33} \\ B_{44} \end{bmatrix} u(t) = Ax(t) + Bu(t)$$

$$y(t) = \begin{bmatrix} C_{11} & C_{22} & C_{33} & C_{44} \end{bmatrix} \begin{bmatrix} x_{\mathrm{I}}(t) \\ x_{\mathrm{II}}(t) \\ x_{\mathrm{III}}(t) \\ x_{\mathrm{IV}}(t) \end{bmatrix} \tag{6.45}$$

where $\varepsilon_1 \gg \varepsilon_2 \gg \varepsilon_3 > 0$ are small positive singular perturbation parameters, $x_{\mathrm{I}}(t) \in \mathrm{R}^{n_1}$ are very slow state variables, $x_{\mathrm{II}}(t) \in \mathrm{R}^{n_2}$ are slow state variables, $x_{\mathrm{III}}(t) \in \mathrm{R}^{n_3}$ are fast state variables, $x_{\mathrm{IV}}(t) \in \mathrm{R}^{n_4}$ are very fast state variables, $n = n_1 + n_2 + n_3 + n_4$, $u(t) \in \mathrm{R}^m$ is the system control input vector, $y(t) \in \mathrm{R}^p$ is the vector of system measurements, and A_{ij}, B_{ii}, and C_{ii}, $i, j = 1, 2, 3, 4,$ are constant matrices of appropriate dimensions. Matrices A_{11},A_{22}, A_{33}, and A_{44} define, respectively, very slow, slow, fast, and very fast subsystem of dimensions n_1, n_2, n_3, and n_4 corresponding to state variables $x_{\mathrm{I}}(t)$, $x_{\mathrm{II}}(t)$, $x_{\mathrm{III}}(t)$, and $x_{\mathrm{IV}}(t)$.

It is a standard assumption in theory of four-time scale linear control systems that the matrices A_{22}, A_{33}, and A_{44} are invertible (Kokotovic et al. 1999; Naidu and Calise 2001), so that the following assumption is imposed for this class of systems:

Assumption 6.4 Matrices A_{22}, A_{33}, and A_{44} are invertible.

In the next section, we will discuss possibilities to implement the four-stage feedback controller design to the four-time scale singularly perturbed system defined in (6.45).

6.4 Future Research Topics

The four-stage controller design can be potentially simplified in the case of four-time scale systems that have a natural decomposition into very slow, slow, fast, and very fast subsystems. Namely, the corresponding design algebraic equations potentially get simpler forms, and due to the presence of three small parameters (that identify the time scales), the L- and P-equations can be efficiently solved, and the existence of their solutions might hold under milder conditions than in the general case.

We have seen in the previous section that *six nonlinear algebraic equations* for L matrices and *six linear Sylvester algebraic equations* for P matrices have to be solved to make the four-stage design feasible. Those equations are summarized here, including the expressions for their coefficients.

The *nonlinear algebraic L-equations* originally derived in (6.4), (6.9), and (6.13) are summarized below. They are given by

$$
\begin{aligned}
f_{41}(L_1, L_2, L_3) &= L_1A_{11} + L_2A_{21} + L_3A_{31} + A_{41} - (L_1A_{14} + L_2A_{24} + L_3A_{34} + A_{44})L_1 = 0\\
f_{42}(L_1, L_2, L_3) &= L_1A_{12} + L_2A_{22} + L_3A_{32} + A_{42} - (L_1A_{14} + L_2A_{24} + L_3A_{34} + A_{44})L_2 = 0\\
f_{43}(L_1, L_2, L_3) &= L_1A_{13} + L_2A_{23} + L_3A_{33} + A_{43} - (L_1A_{14} + L_2A_{24} + L_3A_{34} + A_{44})L_3 = 0
\end{aligned}
\tag{6.46}
$$

$$
\begin{aligned}
f_{31}(L_4, L_5) &= (a_{31} + L_4a_{11} + L_5a_{21}) - (a_{33} + L_4a_{13} + L_5a_{23})L_4 = 0\\
f_{32}(L_4, L_5) &= (a_{32} + L_4a_{12} + L_5a_{22}) - (a_{33} + L_4a_{13} + L_5a_{23})L_5 = 0
\end{aligned}
\tag{6.47}
$$

$$
f_{21}(L_6) = \alpha_{21} + L_6\alpha_{11} - (\alpha_{22} + L_6\alpha_{12})L_6 = 0 \tag{6.48}
$$

The coefficient matrices a_{ij} are defined in (6.6) in terms of solutions of L-equations (6.46), that is, L_1,L_2,L_3, as

$$
\begin{array}{lllll}
a_{11} = A_{11} - A_{14}L_1, & a_{12} = A_{12} - A_{14}L_2, & a_{13} = A_{13} - A_{14}L_3, & a_{14} = A_{14}, & b_{11} = B_{11}\\
a_{21} = A_{21} - A_{24}L_1, & a_{22} = A_{22} - A_{24}L_2, & a_{23} = A_{23} - A_{24}L_3, & a_{24} = A_{24}, & b_{22} = B_{22}\\
a_{31} = A_{31} - A_{34}L_1, & a_{32} = A_{32} - A_{34}L_2, & a_{33} = A_{33} - A_{34}L_3, & a_{34} = A_{34}, & b_{33} = B_{33}
\end{array}
\tag{6.49}
$$

The coefficient matrices α_{ij} are defined in (6.10b) in terms of solutions of (6.47), that is, L_4 and L_5 as

$$\begin{array}{lllll}\alpha_{11} = a_{11} - a_{13}L_4, & \alpha_{12} = a_{12} - a_{13}L_5, & \alpha_{13} = a_{13}, & \alpha_{14} = a_{14}, & \beta_{11} = b_{11} \\ \alpha_{21} = a_{21} - a_{23}L_4, & \alpha_{22} = a_{22} - a_{23}L_5, & \alpha_{23} = a_{23}, & \alpha_{24} = a_{24}, & \beta_{22} = b_{22}\end{array} \tag{6.50}$$

The *six linear Sylvester algebraic P-equations* are defined in (6.23), (6.27), and (6.28), that is

$$\bar{A}_{33}P_3 - P_3(\bar{A}_{44} - \bar{B}_{44}G_4) + \bar{A}_{34} - \bar{B}_{33}G_4 = 0 \tag{6.51}$$

$$\bar{A}_{22}P_{23} - P_{23}\left(\bar{A}_{33} - (\bar{B}_{33} - P_3\bar{B}_{44})G_3\right) + P_{24}\bar{B}_{44}G_3 + \bar{A}_{23} - \bar{B}_{22}G_3 = 0 \tag{6.52}$$

$$\bar{A}_{22}P_{24} - P_{24}(\bar{A}_{44} - \bar{B}_{44}G_4) + \bar{A}_{24} - \bar{B}_{22}G_4 + \bar{A}_{23}P_3 = 0 \tag{6.53}$$

$$\begin{array}{l}S_{11}P_{12} - P_{12}S_{22} - P_{13}S_{32} - P_{14}S_{42} + S_{12} = 0 \\ S_{11}P_{13} - P_{13}S_{33} - P_{14}S_{43} + S_{13} = 0 \\ S_{11}P_{14} - P_{14}S_{44} + S_{14} = 0\end{array} \tag{6.54}$$

The coefficients appearing in the P-equations are derived in (6.3), (6.18), (6.29b), (6.31b), and (6.32b) and given by

$$A_4 = A_{44} + L_1A_{14} + L_2A_{24} + L_3A_{34}, \quad B_4 = B_{44} + L_1B_{11} + L_2B_{22} + L_3B_{33} \tag{6.55}$$

$$\begin{array}{llll}\bar{A}_{11} = \alpha_{11} - \alpha_{12}L_6, & \bar{A}_{12} = \alpha_{12}, & \bar{A}_{13} = \alpha_{13}, & \bar{A}_{14} = \alpha_{14} \\ \bar{A}_{22} = \alpha_{22} + L_6\alpha_{12}, & \bar{A}_{23} = \alpha_{23} + L_6\alpha_{13}, & \bar{A}_{24} = \alpha_{24} + L_6\alpha_{14} & \\ \bar{A}_{33} = \alpha_{33}, & \bar{A}_{34} = \alpha_{34}, & \bar{A}_{44} = a_{44} & \\ \bar{B}_{11} = \beta_{11}, & \bar{B}_{22} = B_{22}, & \bar{B}_{33} = \beta_{33}, & \bar{B}_{44} = B_{44}\end{array} \tag{6.56}$$

$$\bar{B}_{2\xi} = \bar{B}_{22} - P_{23}(\bar{B}_{33} - P_3\bar{B}_{44}) - P_{24}\bar{B}_{44}, \quad \bar{B}_{3\xi} = \bar{B}_{33} - P_3\bar{B}_{44} \tag{6.57}$$

$$S_{14} = \bar{A}_{14} - \bar{B}_{11}G_4 + \bar{A}_{13}P_3 + \bar{A}_{12}P_{24} \tag{6.58}$$

$$\begin{array}{l}S_{11} = \bar{A}_{11}, \quad S_{12} = \bar{A}_{12} - \bar{B}_{11}G_2, \quad S_{13} = \bar{A}_{13} - \bar{B}_{11}G_3 + \bar{A}_{12}P_{23} \\ S_{22} = \bar{A}_{22} - \bar{B}_{2\xi}G_2 \\ S_{32} = -\bar{B}_{3\xi}G_2, \quad S_{33} = \bar{A}_{33} - (\bar{B}_{33} - P_3\bar{B}_{44})G_3 \\ S_{42} = -\bar{B}_{44}G_2, \quad S_{43} = -\bar{B}_{44}G_3, \quad S_{44} = \bar{A}_{44} - \bar{B}_{44}G_4 \\ Q_1 = \bar{B}_{11}, \quad Q_2 = \bar{B}_{22} + \bar{B}_{33} - P_3\bar{B}_{44} - P_{24}\bar{B}_{44}, \quad Q_3 = \bar{B}_{33} - P_3\bar{B}_{44}, \quad Q_4 = \bar{B}_{44}\end{array} \tag{6.59}$$

The assumptions established in Sect. 6.2, as required to guarantee the existence of the solutions of the corresponding algebraic equations and hence to facilitate the four-stage linear feedback controller design are also summarized in this section. The required assumptions are given as follows:

Assumption 6.1 Matrices $\bar{A}_{33}$ and $\bar{A}_{44} - \bar{B}_{44}G_4$ have no eigenvalues in common.

Assumption 6.2a Matrices $\bar{A}_{22}$ and $\bar{A}_{33} - \left(\bar{B}_{33} - P_3\bar{B}_{44}\right)G_3$ have no eigenvalues in common.

Assumption 6.2b The matrix $\bar{A}_{22}$ and matrix $\left(\bar{A}_{44} - \bar{B}_{44}G_4\right)$ have no eigenvalues in common.

Assumption 6.3a Matrices S_{11} and S_{44} have no eigenvalues in common.

Assumption 6.3b Matrices S_{11} and S_{33} have no eigenvalues in common.

Assumption 6.3c Matrices S_{11} and S_{22} have no eigenvalues in common.

As a future research problem, one has to study algebraic equations (6.46, 6.47, 6.48, 6.49, 6.50, 6.51, 6.52, 6.53, 6.54, 6.55, 6.56, 6.57, 6.58, and 6.59) for the structure of the system matrices defined in (6.45), that is, by considering the following changes in (6.46, 6.47, 6.48, 6.49, 6.50, 6.51, 6.52, 6.53, 6.54, 6.55, 6.56, 6.57, 6.58, and 6.59) and Assumptions 6.1–6.3 (note: Assumption 6.4 remains the same since it is specific to the four-time scale linear singularly perturbed system)

$$\begin{array}{llll} A_{11} \rightarrow A_{11}, & A_{12} \rightarrow A_{12}, & A_{13} \rightarrow A_{13}, & A_{14} \rightarrow A_{14} \\ A_{21} \rightarrow \dfrac{1}{\varepsilon_1}A_{21}, & A_{22} \rightarrow \dfrac{1}{\varepsilon_1}A_{22}, & A_{23} \rightarrow \dfrac{1}{\varepsilon_1}A_{23}, & A_{24} \rightarrow \dfrac{1}{\varepsilon_1}A_{24} \\ A_{31} \rightarrow \dfrac{1}{\varepsilon_2}A_{31}, & A_{32} \rightarrow \dfrac{1}{\varepsilon_2}A_{32}, & A_{33} \rightarrow \dfrac{1}{\varepsilon_2}A_{33}, & A_{34} \rightarrow \dfrac{1}{\varepsilon_2}A_{34} \\ A_{41} \rightarrow \dfrac{1}{\varepsilon_3}A_{41}, & A_{42} \rightarrow \dfrac{1}{\varepsilon_3}A_{42}, & A_{43} \rightarrow \dfrac{1}{\varepsilon_3}A_{43}, & A_{44} \rightarrow \dfrac{1}{\varepsilon_2}A_{44} \\ B_{11} \rightarrow B_{11}, & B_{22} \rightarrow \dfrac{1}{\varepsilon_1}B_{22}, & B_{33} \rightarrow \dfrac{1}{\varepsilon_2}B_{33}, & B_{44} \rightarrow \dfrac{1}{\varepsilon_3}B_{44} \end{array} \tag{6.60}$$

and studying them and reformulating the assumptions when the *small* singular perturbation parameters satisfy $\varepsilon_1 \gg \varepsilon_2 \gg \varepsilon_3 > 0$.

Chapter 7
Modeling and System Analysis of PEM Fuel Cells

Fuel cells are electromechanical-electrochemical systems that produce electricity and water from hydrogen and oxygen via the process reverse to water electrolysis. Hydrogen is obtained from hydrogen-rich fuels (natural gas, methanol, ethanol, etc.) using simple chemical-physical processes. Fuel cells produce electricity without burning natural gas (or any other source of hydrogen), so that fuel cells are considered as clean (green) electric energy generators since they do not pollute the environment. A fuel cell is a triode composed of an anode, membrane, and cathode. Hydrogen is pumped from the anode side, and oxygen is pumped from the cathode side. Depending on the type of the membrane, we have several types of fuel cells: proton exchange membrane (polymer electrolyte membrane) (PEM) fuel cells, solid oxide fuel cells, etc. Since the PEM fuel cells are the most developed, best understood, and with numerous applications in the today's world, we will concentrate our attention to this kind of fuel cells.

Modeling, control, and simulation of PEM fuel cells have been a very active research area since the beginning of this century (see, e.g., Padulles et al. (2000), El-Sharkh et al. (2004), Pukrushpan et al. (2004a, b), Fuhrmann et al. (2008), Min et al. (2009), Gou et al. (2010), Kulikovsky (2010), Bavarian et al. (2010), Wang et al. (2011), Becherif et al. (2011), Bareli et al. (2012), Matraji et al. (2013), Bhargav et al. (2014), Eikerling and Kulikovsky (2014), Li et al. (2015a, b), Wu and Zhou (2016), Hong et al. (2017), Tong et al. (2017), Daud et al. (2017), Majlan et al. (2018), and references therein). Importance of mathematical modeling for studying fuel cell dynamics is emphasized in Fuhrmann et al. (2008): "The operation of fuel cells with polymer electrolyte membranes (PEMs) is based on complex interactions of physical, chemical, and electrochemical processes on multiple time scales. A quantitative and qualitative understanding of this complex matter is possible only on the base of mathematical models." Mathematical modeling of fuel cells should be done very carefully combining knowledge from several scientific and engineering disciplines such as mathematics, physics, chemistry, system anal-

V. Radisavljević-Gajić et al., *Multi-Stage and Multi-Time Scale Feedback Control of Linear Systems with Applications to Fuel Cells*, Mechanical Engineering Series,
https://doi.org/10.1007/978-3-030-10389-7_7

ysis, chemical engineering, electrical engineering, mechanical engineering, and in general control systems engineering.

There are several mathematical models of PEM fuel cells that can be found in the scientific and engineering literature. In this chapter, we will review the most standard mathematical models of PEM fuel cells and perform system analysis for some of them starting with the simplest third-order linear and bilinear mathematical models and go to the more complex fifth- and eight-order mathematical models. Modeling of fuel cells is, in general, a pretty complex process due to its multidisciplinary nature that involves various electrochemical, thermodynamic, hydrodynamic, pneumatic, and in general chemical, electrical, and mechanical processes. For simplicity, all gases are assumed to be ideal, but despite this assumption, we will see that the model developed for a PEMFC has many variables and many constants and parameters.

The first mathematical models of fuel cells were derived for three fundamental fuel cell variables: hydrogen pressure, oxygen pressure, and the cathode side water vapor pressure. The models were simple: linear model for PEMFC (El-Sharkh et al. 2004) and bilinear model for PEMFC (Gemmen 2003; Chiu et al. 2004; see also Na et al. 2007). The main advantages of those models were in their simplicity so they can be used for some simple and fast conclusions and calculations about the fuel cell dynamic response, but these models were not good from the control system point of view. It was indicated in Radisavljevic (2011) that in the model of El-Sharkh et al. (2004), the water vapor on the cathode side is an uncontrollable variable. Similarly, it was shown in Radisavljevic-Gajic and Graham (2017) that the bilinear model of Gemmen (2003) and Chiu et al. (2004), linearized at its common operating points, has uncontrollable hydrogen pressure, which limits potential application of this model.

The PEMFC control-oriented model used for a vehicle was developed by Pukrushpan et al. (2004a, b). The fuel cell by itself was modeled as a fifth-order nonlinear system that in addition included dynamics of intake and outtake manifolds, bringing the linearized model dynamics to order eight. A simplified nonlinear third-order model specialized for automotive applications has been developed by Haddad et al. (2015). Another control-oriented fifth-order nonlinear model of a PEMFC was derived in Na and Gou (2008) (see also Gou et al. 2010). There are other recent models of PEMFC that include modeling of humidifiers, water separators, water coolers, and cathode humidity (see, e.g., Rojas et al. (2017), Barzegari et al. (2016), Headley et al. (2016), Chakraborty (2018), Majlan et al. (2018)).

In Sect. 7.1, we first show that the *linear models* of the proton exchange membrane and solid oxide (SO) fuel cells, commonly used in power and energy literature, are not controllable. The source of uncontrollability is the equation for pressure of the water vapor that is only affected by the fuel cell current, which in fact is a disturbance in this system and cannot be controlled by the given model inputs: inlet molar flow rates of hydrogen and oxygen. Being uncontrollable, these models are not good candidates for studying control of dynamic processes in PEM and SO fuel cells. However, due to their simplicity, they can be used in hybrid configurations with other energy-producing devices such as photovoltaic (solar) cells, wind turbines, micro gas turbines, and batteries (ultra-capacitors) to demonstrate some other phenomena, but not for control purposes unless the hybrid models, formed in such hybrid configurations, are controllable. Testing controllability of such hybrid models is mandatory. Secondly, we introduce some algebraic constraints that follow from the model dynamics and the Nernst open-loop fuel cell voltage formula. These constraints must be satisfied in simulation of considered fuel cell modes, for example, via MATLAB/Simulink or any other computer software package.

In Sect. 7.2, we analyze the steady-state dynamic behavior of *a third-order bilinear (nonlinear) model* of a PEM fuel cell. This model is used in several theoretical studies and application papers of PEM fuel cells. We indicate limitations and discuss potential constraints of this mathematical model. We establish conditions for the asymptotic stability at steady state by using the first stability method of Lyapunov. We find that the linearized model at steady state is uncontrollable. Specifically, the state variable corresponding to the hydrogen pressure is not controllable. This means that dynamic deviations of the state space variable corresponding to the hydrogen pressure around the steady-state equilibrium point cannot be controlled. However, due to its stability, the hydrogen pressure as time goes by will tend to the equilibrium point according to its internal (uncontrolled) dynamics so that the model is still applicable for theoretical and practical studies.

In Sect. 7.3, we present the model development for a PEM fuel cell used in the corresponding laboratory at Villanova University. The central part of the laboratory is the Greenlight Innovation G60 Testing Station with TP50 fuel cell. The G60 testing station is designed and built as sophisticated, user customizable, and flexible low-flow testing station for PEM fuel cells that is equipped with Greenlight Innovation TP50 Fuel Cell. The derived model of TP50 Fuel cell in Sect. 7.3.1 is of order five. The state space variables are mass of oxygen in cathode, mass of nitrogen in cathode, mass of hydrogen in anode, mass of water vapor in anode, and mass of water vapor in cathode. The model has two control inputs: mass flow rate of dry hydrogen supply and mass flow rate of dry air supply (oxygen). The produced fuel cell stack current is also an input to the fuel cell. It is treated as a disturbance.

Section 7.4 considers another control-oriented fifth-order nonlinear model of a PEMFC that was derived in Na and Gou (2008) (see also Gou et al. 2010). This model has a similar foundation to the models considered in Purkushpan (2004a, b) and Milanovic et al. (2017) with the different formulation of the control input signals.

In Sect. 7.5, we explain how the eight-order fuel cell model used in electric cars was derived in Pukrushpan et al. (2004a, b). In addition to the five state space variables presented in the model derived in Sect. 7.3, this model has three state space variables that come from the gas dynamics in supply and return manifolds and one from the compressor's blower that provides air (oxygen) to the fuel cell. The linearized model at a nominal operating point and its state space matrices are presented in Sect. 2.4.1.

7.1 Third-Order Linear Model of a PEM Fuel Cell

In the first part of this section, we present a simple third-order linear dynamic model of a PEM fuel cell. Then, we show that this linear model (used also for solid oxide fuel cells) is not controllable. In the follow-up of the section, we introduce some algebraic constraints on this model that follow from system dynamic equations and from the steady-state analysis.

The linear mathematical model for the PEM fuel cell dynamics of three fundamental fuel cell dynamic variables, pressures of hydrogen, oxygen, and water vapor, was derived by El-Sharkh et al. (2004). The model was obtained by keeping the same state equation and slightly modifying the output equation of the mathematical model derived for the SOFC dynamics by Padulles et al. (2000). These linear mathematical models for PEMFC and SOFC have been used in many papers (see, e.g., Uzunoglu and Alam (2006, 2007), Uzunoglu et al. (2007), El-Sharkh et al. (2007), Onar et al. (2010), Yalcinoz et al. (2010), and Wang et al. (2010) for PEMFC-related problems and Zhu and Tomsovic (2002), Li and Rajakaruna (2005), and Hajizadeh and Golkar (2010) for SOFC-related problems).

The linear third-order system state space model by El-Sharkh et al. (2004) was defined by

$$\begin{aligned}\frac{dx_1(t)}{dt} &= -\frac{RTK_{H_2}}{V_A}x_1(t) + \frac{RT}{V_A}q_{H_2}^{in}(t) - \frac{2RTK_r}{V_A}I(t)\\ &= -\frac{1}{\tau_{H_2}}x_1(t) + \frac{1}{\tau_{H_2}K_{H_2}}q_{H_2}^{in}(t) - \frac{2K_r}{\tau_{H_2}K_{H_2}}I(t)\end{aligned} \tag{7.1a}$$

$$\begin{aligned}\frac{dx_2(t)}{dt} &= -\frac{RT}{V_C}K_{O_2}x_2(t) + \frac{RT}{V_C}q_{O_2}^{in}(t) - \frac{RTK_r}{V_C}I(t)\\ &= -\frac{1}{\tau_{O_2}}x_2(t) + \frac{1}{\tau_{O_2}K_{O_2}}q_{O_2}^{in}(t) - \frac{K_r}{\tau_{O_2}K_{O_2}}I(t)\end{aligned} \tag{7.1b}$$

$$\frac{dx_3(t)}{dt} = -\frac{RT}{V_C}K_{H_2O}x_3(t) + \frac{2RTK_r}{V_C}I(t) = -\frac{1}{\tau_{H_2O}}x_3(t) + \frac{2K_r}{\tau_{H_2O}K_{H_2O}}I(t) \tag{7.1c}$$

with the state space variables representing, respectively, hydrogen pressure, oxygen pressure, and the cathode side water vapor pressure

$$x(t) = [x_1(t)\ \ x_2(t)\ \ x_3(t)]^T = \left[p_{H_2}(t)\quad p_{O_2}(t)\quad p_{H_2O}(t)\right]^T \tag{7.2}$$

The output equation represents the measured fuel cell voltage, and it is obtained using the Nernst formula for the open-loop cell voltage, $V_0(t)$, and subtracting losses due to the cell activation, $V_{act}(t)$, and due to the stack (fuel cell) resistance, $V_{ohm}(t)$

$$
\begin{aligned}
y_{PEM}(t) &= V_{PEM}(t) = V_0(t) - V_{act}(t) - V_{ohm}(t) \\
&= N\left(E_0 + \frac{RT}{2F}\ln\left\{\frac{x_1(t)(x_2(t))^{0.5}}{x_3(t)}\right\}\right) - B\ln\left(CI(t)\right) - R^{\text{int}}I(t)
\end{aligned}
\tag{7.3}
$$

where $B = 47.77$ mV and $C = 0.0136\ \text{A}^{-1}$. The system inputs are molar flow rates of hydrogen and oxygen, that is, $q_{H_2}(t)$ and $q_{O_2}(t)$ that can be regulated (controlled). The stack current *I*(*t*) plays a role of a disturbance. Note that *CI*(*t*) must be greater than 1, otherwise the activation voltage will be negative, and hence it will increase the open-loop voltage (instead of reducing it). All other coefficients are assumed to be constant. The values of the constant coefficients defined in the model equations can be found in El-Sharkh et al. (2004).

The SOFC fuel cell model of Padulles et al. (2000) has exactly the same state equations (7.1) and (7.2) but different output equation in which the activation voltage is not present in the expression for the cell output voltage, that is

$$
\begin{aligned}
y_{SO}(t) &= V_{SO}(t)^{SO} = V_{fc}(t) \\
&= N\left(E_0 + \frac{RT}{2F}\ln\left\{\frac{x_1(t)(x_2(t))^{0.5}}{x_3(t)}\right\}\right) - R^{\text{int}}I(t)
\end{aligned}
\tag{7.4}
$$

Of course, in the SOFC mathematical model (7.1), (7.2), and (7.4), the parameters take different values (except for the universal gas constant *R* and the Faraday constant *F*). The values of the constant parameters for the SOFC model defined by (7.1) and (7.4) can be found in (Padulles et al. 2000).

7.1.1 Controllability of the Linear PEM Fuel Cell Model

The controllability and observability concepts are the system state space concepts. They have been known to control engineers for more than 50 years since the initial work of Kalman (1960). Slowly these concepts are becoming known and used in other engineering and scientific disciplines, especially when the so-called Kalman system canonical decomposition was derived in Kalman (1963) (see also Chen 2012). The Kalman canonical decomposition states that only the system modes that are both controllable and observably appear in the system transfer function and those either uncontrollable or unobservable cancel out from the transfer function (system input/output description). This result has established the fact that the state space system description (via system eigenvalues) is more general than the system description via transfer function (via system poles) since the set of system eigenvalues is broader than the set of system poles (all the poles are the eigenvalues, but not all the eigenvalues are the system poles). The importance of controllability in the design of linear controllers for PEM fuel cells was nicely demonstrated in (Serra et al. 2005), where even for originally controllable operating points of a linearized system, some design techniques provide higher controllability measures (requiring less control efforts and more efficient control) than the other also controllable operation points. Controllability analysis of liquid water in a fuel cell has been considered in McCain et al. (2010), where it has been concluded that liquid water controllability is needed to prevent the fuel cell flooding.

In the following, we show that the mathematical models of El-Sharkh et al. (2004) and Padulles et al. (2000) are uncontrollable (zero controllability measure), meaning that no control efforts exist to satisfy general goals of transferring a state variable from a given initial state to a desired final state in a finite time interval Chen (2012). The state space model (7.1) can be represented in the state space form as

$$\begin{aligned}\frac{dx(t)}{dt} &= \begin{bmatrix}\frac{dx_1(t)}{dt}\\ \frac{dx_2(t)}{dt}\\ \frac{dx_3(t)}{dt}\end{bmatrix} = \begin{bmatrix}-\frac{1}{\tau_{H_2}} & 0 & 0\\ 0 & -\frac{1}{\tau_{O_2}} & 0\\ 0 & 0 & 0\end{bmatrix}\begin{bmatrix}x_1(t)\\ x_2(t)\\ x_3(t)\end{bmatrix}\\ &+\begin{bmatrix}\frac{1}{\tau_{H_2}K_{H_2}} & 0\\ 0 & \frac{1}{\tau_{O_2}K_{O_2}}\\ 0 & 0\end{bmatrix}\begin{bmatrix}q_{H_2}^{in}(t)\\ q_{O_2}^{in}(t)\end{bmatrix} + \begin{bmatrix}-\frac{2K_r}{\tau_{H_2}K_{H_2}}\\ -\frac{K_r}{\tau_{O_2}K_{O_2}}\\ \frac{2K_r}{\tau_{H_2O}K_{H_2O}}\end{bmatrix}I(t)\\ &= Ax(t) + Bu(t) + Gd(t)\end{aligned} \tag{7.5}$$

where $u(t) = \begin{bmatrix}q_{H_2}^{in}(t) & q_{O_2}^{in}(t)\end{bmatrix}^T$ are the control inputs and $d(t) = I(t)$ denotes the system disturbance. Using the standard rank controllability test, Chen (2012), we can form the controllability matrix for the state space system model defined in (7.5) as

$$\begin{aligned}\mathrm{C}(A,B) &= \begin{bmatrix}B & AB & A^2B\end{bmatrix}\\ &= \begin{bmatrix}\frac{1}{\tau_{H_2}K_{H_2}} & 0 & -\frac{1}{\tau_{H_2}^2K_{H_2}} & 0 & \frac{1}{\tau_{H_2}^3K_{H_2}} & 0\\ 0 & \frac{1}{\tau_{O_2}K_{O_2}} & 0 & -\frac{1}{\tau_{O_2}^2K_{O_2}} & 0 & \frac{1}{\tau_{O_2}^3K_{O_2}}\\ 0 & 0 & 0 & 0 & 0 & 0\end{bmatrix}\end{aligned} \tag{7.6}$$

It is obvious that the rank of the controllability matrix $\mathrm{C}(A, B)$ is equal to two, that is

$$\mathrm{rank}\{\mathrm{C}(A,B)\} = 2 < 3 = n \tag{7.7}$$

This indicates that in the third-order dimensional linear system considered, only two state variables are controllable and the third one is uncontrollable. Examining the state space equations, it can be observed that the equation for the water vapor is not affected by the control input signal; hence, the water vapor pressure is the uncontrollable variable in this system.

The importance of system controllability in this particular model is needed for several reasons. First of all, it is well known that being uncontrollable, the state variable $x_3(t)$ will even not appear in the system transfer function, Chen (2012). In the case, the transfer function will have the order equal to two corresponding to the controllable state variables $x_1(t)$ and $x_2(t)$. Hence, every frequency domain analysis that involves model (7.1) will be superficial. Secondly, it is known from Chen (2012) that state feedback can be used to stabilize unstable systems, but it cannot make uncontrollable systems controllable so that the variable $x_3(t)$ by no means can be affected by control input signals, and it will remain affected only by the disturbance signal $I(t)$ that changes as $I(t) = V_{fc}(t)/R_L$, where the load R_L changes randomly in time as a piecewise constant. Hence, changes of dynamics of the state variable $x_3(t)$ will be fully determined by its time constant and the fuel cell disturbance (current). It should be emphasized that according to the numerical data from El-Sharkh et al. (2004), the time constant for $x_3(t)$ is much larger than for the remaining two state variables ($\tau_{H_2O} = 18.418\,\text{s}, \tau_{O_2} = 6.64\,\text{s}, \tau_{H_2} = 3.37\,\text{s}$) which means that $x_3(t)$ will take much longer time to reach its steady-state value (when it will be dictated only by the steady-state value of the current $x_3^{ss} = 2K_r I^{ss}/K_{H_2O}$), than the remaining two state variables. Moreover, the magnitude of state variable $x_3(t)$is much smaller than the magnitude of the variables $x_1(t)$ and $x_2(t)$, and it appears in the dominator of the cell output voltage formula (7.4), so that it will have a more dominant, more lasting, and more unpredictable impact on the cell output voltage. Thirdly, controlling water in a fuel cell is fundamentally important (McCain et al. 2010), since it can cause fuel cell flooding, degrade the cell polarization curve, and eventually damage the fuel cell membrane (Pukrushpan et al. 2004a; Nehrir and Wang 2009).

7.1.2 *System Analysis and Constraints of the PEMFC Model*

In this section, we derive some algebraic constraints that follow from the model differential equations. These constraints were not imposed in the papers El-Sharkh et al. (2004) and Padulles et al. (2000) that derived the considered models nor in any other follow-up paper that have used these models alone or in hybrid configurations with other electric energy-generating devices. The constraints are imposed at steady state, for initial conditions, and for all time instants.

Steady-State Constraints

Steady-state analysis of mathematical models for fuel cells is an important technique, as demonstrated by Bavarian et al. (2010). In this subsection, we perform steady-state analysis for the considered third-order linear system.

The steady-state value of the water vapor (obtained by setting the derivative to zero and solving the corresponding algebraic equation, in this case in (7.1c)) is given by

$$x_3^{ss} = \frac{2K_r}{K_{H_2O}} I^{ss} \tag{7.8}$$

The steady-state values of the hydrogen and oxygen pressures are functions of the steady-state values of the molar inlet rates of hydrogen and oxygen and the steady-state fuel cell current. They are, respectively, given by

$$x_1^{ss} = \frac{1}{K_{H_2}} q_{H_2ss}^{in} - \frac{2K_r}{K_{H_2}} I^{ss}, \quad x_2^{ss} = \frac{1}{K_{O_2}} q_{O_2ss}^{in} - \frac{K_r}{K_{O_2}} I^{ss} \tag{7.9}$$

The corresponding steady-state output voltage value (for SOFC) from (7.4) is given by

$$\begin{aligned} y_{SO}^{ss} &= V_{SO}^{ss} = N\left(E_0 + \frac{RT}{2F}\ln\left\{\frac{x_1^{ss}\left(x_2^{ss}\right)^{0.5}}{x_3^{ss}}\right\}\right) - R^{\text{int}} I^{ss} \\ &= N\left(E_0 + \frac{RT}{2F}\ln\left\{\frac{K_{H_2O}}{2K_r K_{H_2}\sqrt{K_{O_2}}} \frac{\left(q_{H_2ss}^{in} - 2K_r I^{ss}\right)\left(q_{O_2ss}^{in} - K_r I^{ss}\right)^{0.5}}{I^{ss}}\right\}\right) - R^{\text{int}} I^{ss} \end{aligned} \tag{7.10}$$

Since the hydrogen and oxygen pressures are positive quantities, the following conditions on the inlet hydrogen and oxygen flow rates must be satisfied at the steady state

$$\begin{aligned} x_1^{ss} &= \frac{1}{K_{H_2}} q_{H_2ss}^{in} - \frac{2K_r}{K_{H_2}} I^{ss} > 0 \quad \Rightarrow \quad q_{H_2ss}^{in} > 2K_r I^{ss} \\ x_2^{ss} &= \frac{1}{K_{O_2}} q_{O_2ss}^{in} - \frac{K_r}{K_{O_2}} I^{ss} > 0 \quad \Rightarrow \quad q_{O_2ss}^{in} > K_r I^{ss} \end{aligned} \tag{7.11}$$

Another set of simulation constraints comes from the Nernst open-loop voltage that is a positive quantity. Since the "ln" operation is present in that formula, we must have

$$\ln\left\{\frac{x_1^{ss}\left(x_2^{ss}\right)^{0.5}}{x_3^{ss}}\right\} > 0 \quad \Rightarrow \quad \frac{x_1^{ss}\left(x_2^{ss}\right)^{0.5}}{x_3^{ss}} > 1 \quad \Rightarrow \quad x_1^{ss}\left(x_2^{ss}\right)^{0.5} > x_3^{ss} \tag{7.12}$$

which implies

$$\left(q^{in}_{\mathrm{H_2}ss} - 2K_r I^{ss}\right)\left(q^{in}_{\mathrm{O_2}ss} - K_r I^{ss}\right)^{0.5} > \frac{2K_r K_{\mathrm{H_2}}\sqrt{K_{\mathrm{O_2}}}}{K_{\mathrm{H_2O}}} I^{ss} \tag{7.13}$$

Hence, the steady-state hydrogen and oxygen pressures must be such to overcome not just the constraint given in (7.11) but also the stronger constraint given in (7.13), which is dictated by the fuel cell (stack) steady-state current limitations.

Initial Conditions and Time Constraints

The constraints analogous to the steady-state constraints imposed in (7.12) and (7.13) must be extended for all time instances. This is due to the fact that pressures are positive quantities at all times, that is, $x_1(t) > 0$, $x_2(t) > 0$, $x_3(t) > 0$, and that from the Nernst formula, we must have

$$\frac{x_1(t)(x_2(t))^{0.5}}{x_3(t)} > 1, \quad \forall t \quad \Rightarrow \quad x_1(t)(x_2(t))^{0.5} > x_3(t), \quad \forall t \tag{7.14}$$

The pressure positivity requirements will also require imposing constraints on the hydrogen and oxygen pressure at the initial time providing the initial condition constraints.

The analytical expression for the hydrogen and oxygen pressures and corresponding initial condition constraints can be obtained by solving the original differential equations given in (7.1)

$$\begin{aligned} x_1(t) &= e^{-\frac{1}{\tau_{\mathrm{H_2}}}t} x_1(0) + \int_0^t e^{-\frac{1}{\tau_{\mathrm{H_2}}}(t-\tau)} \left(q^{in}_{\mathrm{H_2}}(\tau) - \frac{2K_r}{\tau_{\mathrm{H_2}} K_{\mathrm{H_2}}} I(\tau)\right) d\tau > 0, \forall t \\ \Rightarrow \quad x_1(0) &> -\int_0^t e^{\frac{1}{\tau_{\mathrm{H_2}}}\tau} \left(q^{in}_{\mathrm{H_2}}(\tau) - \frac{2K_r}{\tau_{\mathrm{H_2}} K_{\mathrm{H_2}}} I(\tau)\right) d\tau \end{aligned} \tag{7.15}$$

and

$$\begin{aligned} x_2(t) &= e^{-\frac{1}{\tau_{\mathrm{O_2}}}t} x_2(0) + \int_0^t e^{-\frac{1}{\tau_{\mathrm{O_2}}}(t-\tau)} \left(q^{in}_{\mathrm{O_2}}(\tau) - \frac{K_r}{\tau_{\mathrm{O_2}} K_{\mathrm{O_2}}} I(\tau)\right) d\tau > 0, \forall t \\ \Rightarrow \quad x_2(0) &> -\int_0^t e^{\frac{1}{\tau_{\mathrm{O_2}}}\tau} \left(q^{in}_{\mathrm{O_2}}(\tau) - \frac{K_r}{\tau_{\mathrm{O_2}} K_{\mathrm{O_2}}} I(\tau)\right) d\tau \end{aligned} \tag{7.16}$$

It is interesting to observe the initial condition constraints on $x_1(0)$ and $x_2(0)$ imposed in (7.15) and (7.16). Even though the coefficients that multiply the stack current in

(7.15) and (7.16) are small (in El-Sharkh (2004), they are 0.02 and 0.01, respectively, for hydrogen and oxygen), these initial condition constraints must be satisfied in simulation studies whenever these models are used; otherwise in the initial time interval, the negative values could be obtained for the hydrogen and/or oxygen pressures (using MATLAB/Simulink or any other simulation software package). There is no similar initial condition constraint on the water vapor, except for the obvious one that comes from the Nernst formula $x_3(0) \neq 0$ (also, the same formula requires, in general, that $x_3(t) \neq 0, \forall t$). The water vapor values in time as a function of the fuel cell current are obtained by solving the last equation in (7.1), leading to

$$x_3(t) = e^{-\frac{1}{\tau_{H_2O}}t} x_3(0) + \frac{2K_r}{\tau_{H_2O}K_{H_2O}} \int_0^t e^{-\frac{1}{\tau_{H_2O}}(t-\tau)} I(\tau)d\tau \tag{7.17}$$

clearly indicating that the expression for the water pressure is not a function of the inlet molar flow rates of the hydrogen and oxygen.

Since the fuel cell current depends on the load R_L that changes in time independently outside of the fuel cell (as a piecewise constant function), with $V_{fc}(t) = R_L I(t)$, the cell current in general varies, and the above stated constraints, (7.11) and (7.13, 7.14, 7.15 and 7.16), must hold for the worst case scenario (when $I(t)$ takes its maximal values, i.e., for $I_{max} = \max\{I(t)\}, \forall t$).

The lack of controllability of considered linear models of PEM and SO fuel cells might mean that the results presented in journal and conference papers using these uncontrollable models might be valid only for the chosen set of data and that the conclusions drawn in those papers are not general (valid for all possible inputs and all possible values of the state space variables). We have also introduced some algebraic constraints that must be satisfied at steady state, initial time, and at all times to be able to run simulations (in MATLAB/Simulink or any other computer software package) of the considered linear PEM and SO fuel cell models.

7.2 Third-Order Bilinear PEM Fuel Cell Model

The bilinear (nonlinear) third-order mathematical model of PEM fuel cell that considers dynamics of three fundamental variables, pressures of hydrogen, oxygen, and water vapor, was derived by Gemmen (2003) and Chiu et al. (2004). This model was developed in the course of research conducted at the US Department of Energy National Energy Technology Laboratory at the University of West Virginia, Morgantown (Chiu et al. 2004; Page et al. 2007; Na et al. 2007; Gou et al. 2010).

In this section, we present system analysis and establish steady-state constraints and limitations of the PEM fuel cell nonlinear mathematical model derived and considered by Gemmen (2003) and Chiu et al. (2004). Controlling and observing dynamics of processes in fuel cell require that the corresponding mathematical models be controllable and observable or at least stabilizable (unstable state variables are controllable) and detectable (unstable state variables are observable) (Chen 2012).

Before designing controllers for dynamic systems for which mathematical models are available, one should first perform system analysis of such models, study corresponding transient and steady-state limitations, establish design conditions and constraints, simulate such models with and without corresponding controllers, and implement the actual controllers to the real physical systems, in this case fuel cells.

The state space variables of the bilinear mathematical model of Gemmen (2003) and Chiu et al. (2004) are pressures of hydrogen, oxygen, and water vapor, that is

$$x(t) = \begin{bmatrix} x_1(t) & x_2(t) & x_3(t) \end{bmatrix}^T = \begin{bmatrix} p_{\mathrm{H_2}}(t) & p_{\mathrm{O_2}}(t) & p_{\mathrm{H_2O}}(t) \end{bmatrix}^T \tag{7.18}$$

The model input variables are the molar flow rates of hydrogen $W_{\mathrm{H_2}}^{in}(t)$ (pumped from the anode side), oxygen $W_{\mathrm{O_2}}^{in}(t)$, and water vapor $W_{\mathrm{H_2O}}^{in}(t)$ (both pumped from the cathode side; the water vapor is needed to humidify the cell membrane; moreover, the PEM fuel cell produces water on the cathode side)

$$\begin{bmatrix} u_1(t) & u_2(t) & u_3(t) \end{bmatrix}^T = \begin{bmatrix} W_{\mathrm{H_2}}^{in}(t) & W_{\mathrm{O_2}}^{in}(t) & W_{\mathrm{H_2O}}^{in}(t) \end{bmatrix} \tag{7.19}$$

The state space differential equations of the considered fuel cell mathematical model are

$$\frac{dx_1(t)}{dt} = \frac{RT}{V_A}\left(1 - \frac{x_1(t)}{p_{op}}\right)u_1(t) + \frac{RT}{V_A}\left(\frac{x_1(t)}{p_{op}} - 1\right)2K_rI(t) = f_1(x,u,I) \tag{7.20}$$

$$\begin{aligned} \frac{dx_2(t)}{dt} &= \frac{RT}{V_C}\left(1 - \frac{x_2(t)}{p_{op}}\right)u_2(t) + \frac{RT}{V_C}\left(\frac{x_2(t)}{p_{op}} - 1\right)K_rI(t) - \frac{RT}{V_C}u_3(t)\frac{x_2(t)}{p_{op}} \\ &= f_2(x,u,I) \end{aligned} \tag{7.21}$$

$$\begin{aligned} \frac{dx_3(t)}{dt} = \frac{RT}{V_C}\left(1 - \frac{x_3(t)}{p_{op}}\right)u_3(t) + \frac{RT}{V_C}\left(1 - \frac{x_3(t)}{p_{op}}\right)2K_rI(t) \\ - \frac{RT}{V_C}u_2(t)\frac{x_3(t)}{p_{op}} = f_3(x,u,I) \end{aligned} \tag{7.22}$$

The cell output voltage is given by the well-known Nernst formula as

$$y(t) = V(t) = N\left(E_0 + \frac{RT}{2F}\ln\left\{\frac{x_1(t)(x_2(t)/P_{std})^{0.5}}{x_3(t)}\right\} - L(t)\right) \\ = y(x_1(t), x_2(t), x_3(t)) \tag{7.23}$$

In (7.20, 7.21 and 7.22), $I(t)$ stands for the cell current that is considered as a system disturbance since it changes as $I(t) = V(t)/R_L$, where R_L representing the load is a random quantity. In formula (7.23), $L(t)$ represents the voltage losses due to fuel cell activation, electrical resistance, and the concentration loss. The constant quantities used in (7.20, 7.21, 7.22 and 7.23) are the following: R is the universal gas constant, T is temperature, V_A and V_C are anode and cathode volumes, p_{op} is the operating pressure (steady-state anode and cathode pressures), F is the Faraday constant, N represents the number of fuel cells, $K_r = N/4F$ is the given constant, and P_{std} is the standard pressure.

The dynamic system defined by (7.20, 7.21 and 7.22) represents a *bilinear system* (the first class of nonlinear systems) since it contains as nonlinearities only the products of state and control variables. The system output defined in (7.23) is nonlinear. This model was derived based on the assumption that the anode and cathode pressures remain constant at steady state and equal to the operating pressure p_{op} and that the PEM fuel cell temperature is constant. The model is expected to perform well at steady state. In this section, we will be concerned with its steady-state system analysis and provide additional results on the model stability, controllability, and observability, as well as established algebraic constraints among state space variables that have to be satisfied at steady state to guarantee system stability of this fuel cell model.

The considered bilinear (nonlinear) PEM fuel cell model is convenient due to its simplicity. It is only of order three and provides information about dynamic changes of three most important fuel cell variables: pressures of hydrogen, oxygen, and water vapor. It is important to emphasize that other nonlinear PEM fuel cell models of order three exist in the literature (see, e.g., Chen 2013). They provide information about different state variables. For example, Chen (2013) chooses the state space variables pressures of hydrogen, oxygen, and activation voltage and derives the corresponding third-order nonlinear model.

7.2.1 Steady-State PEM Fuel Cell Equilibrium Points

General analysis and design of a PEM fuel cell was considered in Pandiyan et al. (2013). Steady-state analysis of mathematical models for fuel cells is an important technique that brings more light into dynamical behavior of important state space variables of considered fuel cell models (Bavarian et al. 2010) and establishes model limitations and constraints (Chakraborty 2018).

In this subsection, we perform steady-state analysis for the considered third-order linear system. We specialize our analysis to the PEM fuel cell model of Gemmen

(2003) and Chiu et al. (2004) presented in (7.18, 7.19, 7.20, 7.21, 7.22, 7.23, 7.24, 7.25, 7.26, 7.27, 7.28, 7.29, 7.30, 7.31, 7.32, 7.33 and 7.34). Assuming that the fuel cell model in (7.20, 7.21 and 7.22) is asymptotically stable and that the fuel cell inputs and the fuel cell current are constant in a sufficiently long period of time, then the system state variables will reach their constant steady state values. The steady-state values of the fuel cell model defined in (7.20, 7.21 and 7.22) are obtained from

$$0 = \alpha_A \left(1 - \frac{x_{1ss}}{p_{op}}\right) u_{1ss} + \alpha_A \left(\frac{x_{1ss}}{p_{op}} - 1\right) 2K_r I_{ss} \tag{7.24}$$

$$0 = \alpha_C \left(1 - \frac{x_{2ss}}{p_{op}}\right) u_{2ss} + \alpha_C \left(\frac{x_{2ss}}{p_{op}} - 1\right) K_r I_{ss} - \alpha_C \frac{x_{2ss}}{p_{op}} u_{3ss} \tag{7.25}$$

$$0 = \alpha_C \left(1 - \frac{x_{3ss}}{p_{op}}\right) u_{3ss} + \alpha_C \left(1 - \frac{x_{3ss}}{p_{op}}\right) 2K_r I_{ss} - \alpha_C \frac{x_{3ss}}{p_{op}} u_{2ss} \tag{7.26}$$

where $\alpha_A = RT/V_A$ and $\alpha_C = RT/V_C$. The algebraic equations (7.24, 7.25 and 7.26) lead to the steady-state relations

$$0 = \left(p_{op} - x_{1ss}\right)\left(u_{1ss} - 2K_r I_{ss}\right) \tag{7.27}$$

$$0 = \left(p_{op} - x_{2ss}\right)\left(u_{2ss} - K_r I_{ss}\right) - x_{2ss} u_{3ss} \tag{7.28}$$

$$0 = \left(p_{op} - x_{3ss}\right)\left(u_{3ss} + 2K_r I_{ss}\right) - x_{3ss} u_{2ss} \tag{7.29}$$

Mathematical solutions of algebraic equation (7.27) give the following possibilities for the steady-state values of the hydrogen pressure and the hydrogen inlet flow molar rates:

$$x_{1ss} = p_{op} \text{ and } u_{1ss} = \text{anything, including } u_{1ss} = 2K_r I_{ss} \tag{7.30a}$$

$$u_{1ss} = 2K_r I_{ss} \text{ and } x_{1ss} = \text{anything, including } x_{1ss} = p_{op} \tag{7.30b}$$

It is interesting to observe that the MATLAB/Simulink simulation result obtained using the data from Chiu et al. (2004) has produced the solution given in (7.30). As a matter of fact, we will show in the stability analysis presented in the rest of this section that in order to guarantee stability at steady state, the following condition must be satisfied $u_{1ss} > 2K_r I_{ss}$. The solution (7.30) is ruled out since it does not satisfy the fuel cell stability condition. The way the original DoE model is defined in Chiu et al. (2004), "it is assumed that the cell anode pressures remain constant and equal at p_{op}, which is the steady state operating pressure." Hence, a constraint is imposed that at steady state the anode pressure ($p_A(t) = p_{H_2}(t) = x_1(t) \to x_{1ss}$, for $t \to \infty$) is equal to the cathode pressure at steady state, which produces

$$p_C(t) = p_{O_2}(t) + p_{H_2O} = x_2(t) + x_3(t) \to x_{2ss} + x_{3ss}, \text{for } t \to \infty$$

and

$$p_{Ass} = x_{1ss} = p_{op} = p_{Css} = x_{2ss} + x_{3ss} = p_{op} \tag{7.31}$$

It should be noted also that in practice $x_{2ss} >> x_{3ss}$.

In the following, we study the impact of the derived steady-state conditions on the fuel cell model stability, controllability, and observability, namely, we will examine the dynamic behavior of the nonlinear fuel cell model defined by (7.20, 7.21 and 7.22) around its equilibrium point defined by its steady-state values and study stability, controllability, and observability at that point.

7.2.2 *Fuel Cell System Stability Analysis*

Stability of the PEM fuel cell model (7.20, 7.21 and 7.22) can be determined using the first method of Lyapunov, Khalil (2002). To that end, we linearize the state equations (7.20, 7.21 and 7.22) and evaluate the linearized system state space matrix at the steady-state point and check the location of its eigenvalues. Since we will study in the next section the system controllability at steady state, we will derive here also the input matrix of the linearized system. The linearized system is defined by

$$\frac{d\Delta x(t)}{dt} = A\Delta x(t) + B\Delta u(t) \tag{7.32}$$

where

$$\begin{gathered} x_i(t) = x_{iss} + \Delta x_i(t), u_i(t) = u_{iss} + \Delta u_i(t), \quad i = 1,2,3 \\ \Delta x(t) = [\,\Delta x_1(t) \quad \Delta x_2(t) \quad \Delta x_3(t)\,]^T,\ \Delta u(t) = [\,\Delta u_1(t) \quad \Delta u_2(t) \quad \Delta u_3(t)\,]^T \end{gathered} \tag{7.33}$$

We first find the Jacobian matrices A and B of (7.32) and evaluate them at the steady state points, which leads to

$$A = \begin{bmatrix} \frac{\partial f_1}{\partial x_1} & \frac{\partial f_1}{\partial x_2} & \frac{\partial f_1}{\partial x_3} \\ \frac{\partial f_2}{\partial x_1} & \frac{\partial f_2}{\partial x_2} & \frac{\partial f_2}{\partial x_3} \\ \frac{\partial f_3}{\partial x_1} & \frac{\partial f_3}{\partial x_2} & \frac{\partial f_3}{\partial x_3} \end{bmatrix}_{|ss} = \begin{bmatrix} a_{11} & 0 & 0 \\ 0 & a_{22} & 0 \\ 0 & 0 & a_{33} \end{bmatrix} \tag{7.34a}$$

where

$$a_{11} = \frac{\alpha_A(2K_rI_{ss} - u_{1ss})}{p_{op}}, \quad a_{22} = \frac{\alpha_C(K_rI_{ss} - u_{2ss} - u_{3ss})}{p_{op}}$$
$$a_{33} = -\frac{\alpha_C(2K_rI_{ss} + u_{2ss} + u_{3ss})}{p_{op}} \tag{7.34b}$$

and

$$B = \begin{bmatrix} \frac{\partial f_1}{\partial u_1} & \frac{\partial f_1}{\partial u_2} & \frac{\partial f_1}{\partial u_3} \\ \frac{\partial f_2}{\partial u_1} & \frac{\partial f_2}{\partial u_2} & \frac{\partial f_2}{\partial u_3} \\ \frac{\partial f_3}{\partial u_1} & \frac{\partial f_3}{\partial u_2} & \frac{\partial f_3}{\partial u_3} \end{bmatrix}_{|ss} \tag{7.35}$$

$$B = \frac{1}{p_{op}} \begin{bmatrix} \alpha_A(p_{op} - x_{1ss}) & 0 & 0 \\ 0 & \alpha_C(p_{op} - x_{2ss}) & -\alpha_C x_{2ss} \\ 0 & -\alpha_C x_{3ss} & \alpha_C(p_{op} - x_{3ss}) \end{bmatrix}$$

It can be seen from (7.34) that the eigenvalues of the matrix A will be all in the left half complex plane (providing linear dynamic system asymptotic stability in the sense of Lyapunov, Khalil (2002) if the following conditions are simultaneously met:

$$u_{1ss} > 2K_rI_{ss} \tag{7.36}$$

and

$$u_{2ss} + u_{3ss} > K_rI_{ss} \tag{7.37}$$

From conditions (7.30) and (7.31), it follows that at the stable equilibrium point, the following must hold:

$$x_{1ss} = p_{op} \tag{7.38}$$

Note that the derivations presented in (7.24, 7.25, 7.26, 7.27, 7.28, 7.29, 7.30, 7.31, 7.32, 7.33, 7.34, 7.35, 7.36, 7.37 and 7.38) are the same for a single cell and the fuel cell stack composed of N single fuel cells placed in series since the steady-state current I_{ss} is the same for all cells. Due to the facts that u_{1ss} and u_{2ss} or u_{3ss} may take arbitrarily constant values, the bilinear models might not have the same steady-state equilibrium points, but all of them will have the same current.

7.2.3 PEM Fuel Cell Controllability and Observability Analysis

The controllability and observability conditions have become particularly important for optimal control and estimation (observation) problems associated with fuel cells, as demonstrated in several papers (Serra et al. 2005; McCain et al. 2010; Radisavljevic 2011).

The conditions (7.31) and (7.38) imply that the linearized input matrix at the asymptotically stable equilibrium point is given by

$$B = \frac{1}{p_{op}} \begin{bmatrix} 0 & 0 & 0 \\ 0 & \alpha_C\left(p_{op} - x_{2ss}\right) & -\alpha_C x_{2ss} \\ 0 & -\alpha_C x_{3ss} & \alpha_C\left(p_{op} - x_{3ss}\right) \end{bmatrix} = \begin{bmatrix} 0 & 0 & 0 \\ 0 & b_{22} & b_{23} \\ 0 & -b_{22} & -b_{23} \end{bmatrix} \tag{7.39}$$

Using the standard controllability test (Chen 2012), we can form the controllability matrix for the state space system defined in (7.32) at the equilibrium point and examine its rank (that is supposed to be equal to three (the order of the system) for system controllability). It can be shown that for the considered third-order system, we have

$$\mathrm{rank}\{\mathcal{C}(A,B)\} = \mathrm{rank}\{\begin{bmatrix} B & AB & A^2B \end{bmatrix}\} = \mathrm{rank}[\mathcal{C}_1 \quad \mathcal{C}_2] \le 2 \tag{7.40}$$

$$\mathcal{C}_1 = \begin{bmatrix} 0 & 0 & 0 \\ b_{22} & b_{23} & a_{22}b_{22} \\ -b_{22} & -b_{23} & -a_{33}b_{22} \end{bmatrix} \tag{7.41a}$$

$$\mathcal{C}_2 = \begin{bmatrix} 0 & 0 & 0 \\ a_{22}b_{23} & a_{22}^2 b_{22} & a_{22}^2 b_{23} \\ -a_{33}b_{23} & -a_{22}^2 b_{22} & -a_{22}^2 b_{23} \end{bmatrix} \tag{7.41b}$$

which clearly indicates that the considered *fuel cell system mathematical model is uncontrollable at steady state*. The fact that the rank of the controllability matrix is equal to two indicates that this system has two controllable and one uncontrollable state space variable.

Even more, using the eigenvalue Popov-Belevitch-Hautus controllability test (Chen 2012), we are able to identify which system mode (eigenvalue or state space variable) is not controllable. Since the matrix A of the linearized system given in (7.32) is diagonal, its eigenvalues are identical to its diagonal elements and given by

$$\lambda_1 = a_{11}, \quad \lambda_2 = a_{22}, \quad \lambda_3 = a_{33} \tag{7.42}$$

We examine the rank of matrices

$$\text{Rank}\{[\lambda_i I - A \quad B]\}, \quad i = 1, 2, 3 \tag{7.43}$$

that must be equal to three for a particular eigenvalue to be controllable, according to the Popov-Belevitch-Hautus test (Chen 2012). We found that this rank is equal to two for the eigenvalue λ_1, namely

$$\text{Rank}\left\{\begin{bmatrix} 0 & 0 & 0 & 0 & 0 & 0 \\ 0 & \lambda_1 - \lambda_2 & 0 & 0 & b_{22} & b_{23} \\ 0 & 0 & \lambda_1 - \lambda_3 & 0 & -b_{22} & -b_{23} \end{bmatrix}\right\} = 2 \tag{7.44}$$

This indicates that the first state space variable that corresponds to the hydrogen pressure is not controllable, which implies that this state variable (hydrogen pressure) cannot be controlled. However, due to its stability, the hydrogen pressure will go to the equilibrium point according to its internal (uncontrolled) dynamics so that the model is still applicable for theoretical and practical studies. Since under asymptotic stability conditions imposed in the previous section, the linearized fuel cell model is stabilizable (uncontrollable system modes are asymptotically stable; Chen 2012), which facilitates the design of optimal linear controllers.

To examine observability of the considered fuel cell mathematical model, we find first the output matrix of the linearized fuel cell model at steady state. After some algebra, it can be shown that it is given by

$$\begin{aligned} C &= \begin{bmatrix} \frac{\partial y}{\partial x_1} & \frac{\partial y}{\partial x_2} & \frac{\partial y}{\partial x_2} \end{bmatrix}_{|ss} \\ &= \frac{RT}{4F}\begin{bmatrix} \frac{1}{x_{1ss}} & \frac{1}{2x_{2ss}} & -\frac{1}{x_{3ss}} \end{bmatrix} = \begin{bmatrix} c_{11} & c_{12} & c_{13} \end{bmatrix} \end{aligned} \tag{7.45}$$

This result reveals that all variables of the linearized fuel cell model are observable at steady state since every single state variable is measured directly due to the facts that the linearized system matrix is diagonal and the linearized output matrix has only one row with all non-zero elements. Hence, all three state variables (pressures of hydrogen, oxygen, and water vapor) of the considered linearized fuel cell model can be observed at all times through one-dimensional output equation using linear observers (Sinha 2007; Chen 2012).

Having a stabilizable-observable system *model*, an optimal linear-quadratic controller can be designed (Sinha 2007). To design an optimal linear-quadratic controller for the actual *physical* fuel cell *system*, one will need to construct also an observer

(Sinha 2007) to estimate all three state variables since the linearized measurement formula (7.44) provides only one equation for three unknown state space variables.

7.2.4 Simulation Results

The experimental validation of the mathematical model of the PEM fuel cell considered in this section, and given in formulas (7.18, 7.19, 7.20, 7.21, 7.22 and 7.23), was conducted at the DoE National Energy Technology Laboratory at the University of West Virginia, Morgantown, for which the results "show generally good agreement," as indicated by Chiu et al. (2004) and Gou et al. (2010). Since the purpose of our study is to indicate limitations and constraints of the developed mathematical model of Gemmen (2003), Chiu et al. (2004), and Gou et al. (2010), we perform simulation using MATLAB/Simulink. Validation of results using simulation of mathematical models seems to be a common practice in fuel cell research (Gemmen 2003; Gou et al. 2010; Becherif et al. 2011; Fuhrmann et al. 2008; Radu and Taccani 2006; Abdin et al. 2017). Another way to verify results obtained is to build corresponding electrical circuits, which was done in several papers (Famouri and Gemmen 2003; Page et al. 2007; Arsov 2007; Gou et al. 2010).

We have built the considered fuel cell model MATLAB/Simulink block diagram and presented it in Fig. 7.1.

We use the same data as the ones used by Chiu et al. (2004), that is, $N = 4$, $V_A = 6.495 \times 10^{-6}\,\mathrm{m}^3$, $V_C = 12.96 \times 10^{-6}\,\mathrm{m}^3$, $T = 338.5\,\mathrm{K}$, $F = 96439\,\mathrm{C/mol}$, $P_{std} = 101325\,\mathrm{Pa} = 1\,\mathrm{atm}$, $R = 8.3144\,\mathrm{J/(K \times mol)} = 0.082\,\mathrm{atm \times L/(K \times mol)}$, and $p_{op} = 101000$ Pa. For the input signals, we have also used the same signals by Chiu et al. (2004), that is

$$
\begin{aligned}
u_1 &= 3664\,\mathrm{ml/min} = \left(3664 \times 10^{-3}/60\right) \times p_{op}/(RT)\,\mathrm{mol/s} = 2.1915\,\mathrm{mol/s} \\
u_2 &= 11548\,\mathrm{ml/min} = \left(11548 \times 10^{-3}/60\right) \times p_{op}/(RT)\,\mathrm{mol/s} = 6.9070\,\mathrm{mol/s} \\
u_3 &= 112\,\mathrm{ml/min} = \left(112 \times 10^{-3}/60\right) \times p_{op}/(RT)\,\mathrm{mol/s} = 0.0670\,\mathrm{mol/s}
\end{aligned}
$$

Since we are interested in the dynamic behavior of the fuel cell state space variables in the neighborhood of the steady-state points, we have appropriately chosen the initial conditions as $p_{H_2}(0) = 0.1 p_{op}$, $p_{O_2}(0) = 0.1 p_{op}$, and $p_{H_2O}(0) = 0.01 p_{op}$ where $p_{op} = 101000$ Pa. We have performed also some additional analysis that is needed for the purpose of simulation. Under conditions given in (7.31), that is, $x_{2ss} + x_{3ss} = p_{op}$, the algebraic equations (7.28) and (7.29) for u_{2ss} and u_{3ss} become linearly dependent so that they represent only one algebraic equation. Assuming that u_{2ss} and x_{2ss} or x_{3ss} values are given, then the steady-state value for u_{3ss} satisfies

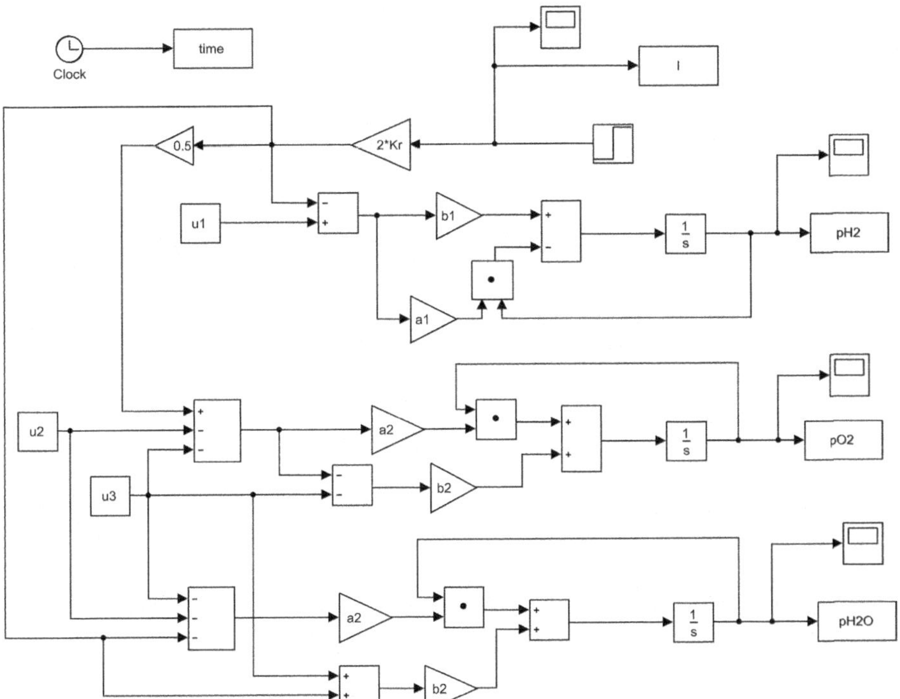

Fig. 7.1 Simulink block diagram for the PEM fuel cell model with $a_1 = \alpha_A/p_{op}$, $a_2 = \alpha_B/p_{op}$, $b_1 = \alpha_A$, $b_2 = \alpha_B$. Integrator initial conditions are $p_{H_2}(0) = 0.9p_{op}$, $p_{O_2}(0) = 0.9p_{op}$, and $p_{H_2O}(0) = 0$

$$u_{3ss} = \frac{1}{x_{2ss}}\left(p_{op} - {}^{op}x_{2ss}\right)\left(u_{2ss} - K_r I_{ss}\right) = \frac{x_{3ss}}{p_{op} - x_{3ss}}\left(u_{2ss} - K_r I_{ss}\right) \tag{7.46}$$

The other way around, u_{3ss} and x_{2ss} or x_{3ss} values given, requires that u_{2ss} satisfies

$$u_{2ss} = \frac{1}{x_{3ss}}\left(p_{op} - {}^{op}x_{3ss}\right)\left(u_{3ss} + 2K_r I_{ss}\right) = \frac{x_{2ss}}{p_{op} - x_{2ss}}\left(u_{3ss} + 2K_r I_{ss}\right) \tag{7.47}$$

For u_{2ss} and u_{3ss} given (which is the case for the constant inputs), the steady-state values for the cathode oxygen and water vapor pressures can be obtained from (7.28) to (7.29), and using the fact that $K_r I_{ss}$ is very small (order of 10^{-4} for $I_{ss} = 40\,\text{A}$), they can be approximated as

$$x_{2ss} = \frac{u_{2ss} - K_r I_{ss}}{u_{2ss} + u_{3ss} - K_r I_{ss}} p_{op} \approx \frac{u_{2ss}}{u_{2ss} + u_{3ss}} p_{op} \tag{7.48}$$

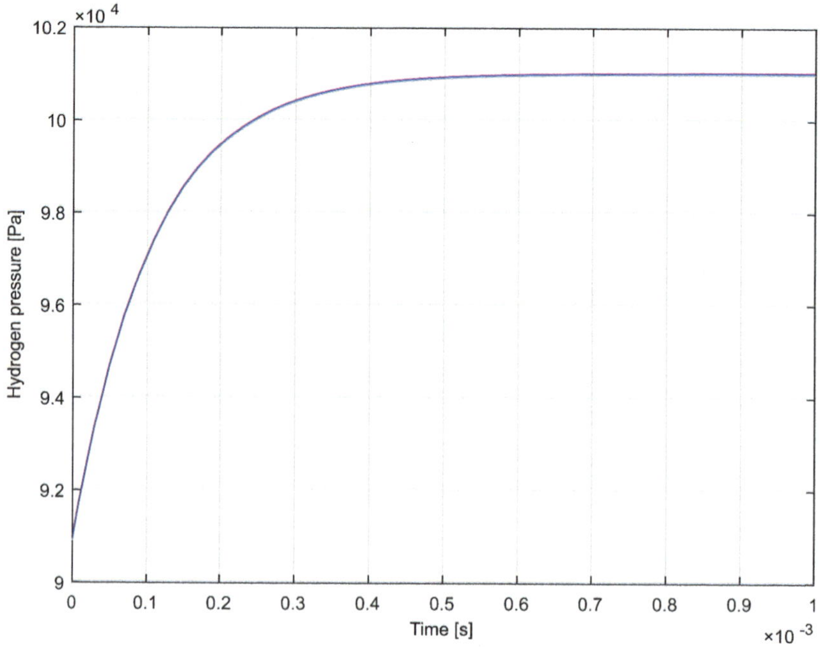

Fig. 7.2 Hydrogen pressure transient and steady responses with $p_{H_2}(0) = 0.9p_{op}$

$$x_{3ss} = \frac{u_{3ss} + 2K_r I_{ss}}{u_{2ss} + u_{3ss} + 2K_r I_{ss}} p_{op} \approx \frac{u_{3ss}}{u_{2ss} + u_{3ss}} p_{op} \tag{7.49}$$

From (7.48) and (7.49), we have an interesting approximate formula for the ratio of the cathode oxygen and water vapor pressures

$$\frac{x_{2ss}}{x_{3ss}} \approx \frac{u_{2ss}}{u_{3ss}} \tag{7.50}$$

Using the given data, we have found values such as $x_{1ss} = p^{ss}_{H_2} = 101,000\,\text{Pa}$, $x_{2ss} = p^{ss}_{O_2} = 100,030\,\text{Pa}$, and $x_{3ss} = p^{ss}_{H_2O} = 982.1\,\text{Pa}$ at the steady state, which produce $\frac{x_{2ss}}{x_{3ss}} = 102.8714$.

Based on the simulation results, we have obtained the following quantity: $x_{1ss} = 101{,}000\,\text{Pa} \approx x_{2ss} + x_{3ss} = 101{,}012\,\text{Pa}$. This indicates that the anode and cathode pressures are pretty well balanced at steady state (i.e., needed to extend the membrane life). Using the approximate formula (7.50), we have $\frac{x_{2ss}}{x_{3ss}} \approx \frac{6.9070}{0.0670} = 103.0896$, which indicates that the derived formula (7.50) represents a very good approximation. In Figs. 7.2, 7.3 and 7.4, we present the fuel cell responses for the considered values of the parameters and to the given inputs.

It can be seen from these figures that it takes less than 1 ms for the variables to reach their constant steady-state values, even though the steady-state values are theoretically obtained for $t \to \infty$. If for some reason, due to disturbances and

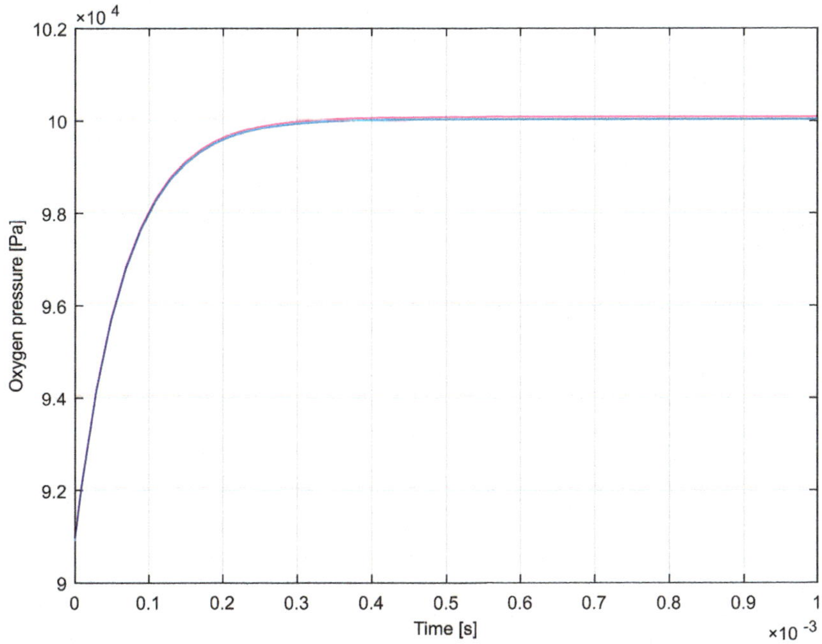

Fig. 7.3 Oxygen pressure transient and steady-state responses staring with $p_{O_2}(0) = 0.9p_{op}$

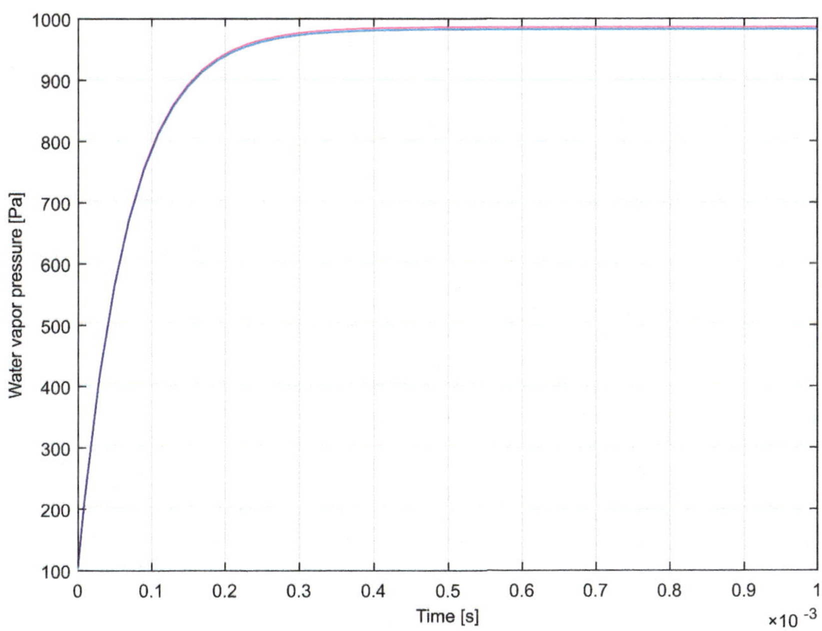

Fig. 7.4 Water vapor pressure transient and steady-state responses, $p_{H_2O}(0) = 0$

inaccuracies, the hydrogen pressure is away from its steady-state value, no hydrogen pressure deviations control signal exists since we have established in the previous section that the hydrogen pressure is an uncontrollable state variable in this fuel cell model.

The asymptotic stability conditions given in formulas (7.36) and (7.37) hold since $2K_rI_{ss} = 8.2964 \times 10^{-4} < u_{1ss} < u_{2ss} + u_{3ss}$. Hence, since the hydrogen pressure at steady state is asymptotically stable, all deviations of the hydrogen pressure around steady-state values will go back to the steady state with its internal dynamics determined by the corresponding time constant of the linearized system given in (7.34) by $1/a_{11} = p_{op}/\alpha_A(2K_rI_{ss} - u_{1ss})$.

The importance of PEM fuel cell modeling and fulfillment of the corresponding physical, chemical, and mathematical constraints should be even more emphasized in cases when PEM fuel cells are coupled to corresponding hydrogen gas reformers (fuel cell processors that produce hydrogen from hydrogen rich fuels) that have their complex multi-time scale dynamics, Ridisavljevic-Gajic and Rose (2014).

If a controller is needed to be designed for the linearized system dynamics of the considered fuel cell model, one will be able to use the linear-quadratic optimal controller since the linearized state space model given by (7.32, 7.33, 7.34, 7.35) and (7.44) is stabilizable-observable (Sinha 2007; Chen 2012). However, the eigenvalue assignment full-state feedback controller (also known as the pole placement controller design technique) cannot be designed due to the lack of system controllability (Sinha 2007; Chen 2012). Since (7.32) and (7.44) represent an observable state space linearized system, if needed, a linearized observer (either full-order or reduced-order) can be designed for the purpose of observing (estimating, monitoring) deviations of all three state space variables around their operating (nominal, steady state) points. Using these estimated state space variables, observer (full-order or reduced-order)-driven feedback optimal controllers can be designed.

It is important to emphasize that when feeding back the signals for control purposes, the unit measure conversion table must be used. For example, in the block diagram presented in Fig. 7.1, the input signals are in mol/s, but the output signals (pressures of oxygen and hydrogen) are in Pa. The feedback control signals must be represented in the same units consistent with the input signal units, that is, in mol/s.

7.3 Greenlight Innovation G60 Station with TP50 PEMFC

Greenlight Innovation G60 is a fully automated testing station system that is equipped with a TP50 fuel cell, also supplied by the Greenlight Innovation.[1] Its specifications are presented in Table 7.1.

[1]Manufacturer's website: www.greenlightinnovation.com

Table 7.1 Greenlight Innovation G60 general specifications

Approximate power range	1–500 W
Gas flows	Mass flow controllers (MFC)
Anode flow range	0.1–10 nlpm
Cathode flow range	0.2–20 nlpm
Gas humidification technology	Contact humidifier
Dew point control	Up to 95 °C
Gas temperature	Up to 110 °C
Cell pressure control	Up to 300 kPa
Load bank (max current, voltage)	60 A, 100 V

nlpm normal liters per minute

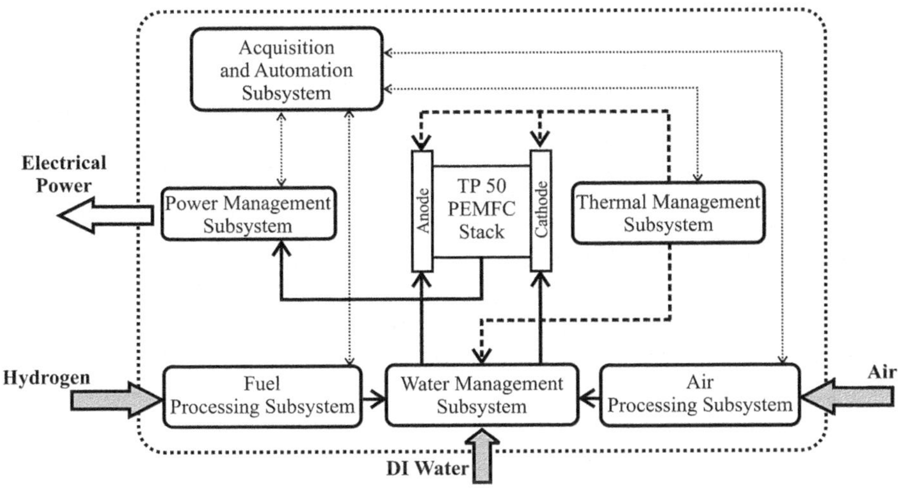

Fig. 7.5 Greenlight Innovation G60 testing station, general schematic

The system can be split into a couple of subsystems that all together manage the operating conditions during the testing. As it can be observed from Fig. 7.5, there are fuel and air (oxygen) processing subsystems, thermal management subsystem, water management subsystem, power conditioning subsystem, and acquisition and automation subsystem. General specifications are listed in Table 7.1.

Fuel/air (oxygen) processing subsystems are responsible for maintaining set point values of mass flows and inlet pressures of the gases. Thermal management subsystem control loops maintain temperatures of TP50 fuel cell, and supply pipes, which are surrounded by a heat tape. Also, thermal management subsystem controls the gas humidifiers that are parts of the water management subsystem. System inputs are dry gases that are led through those humidifiers, which use externally supplied DI water. Simplified schematic of the humidification process is shown in Fig. 7.6.

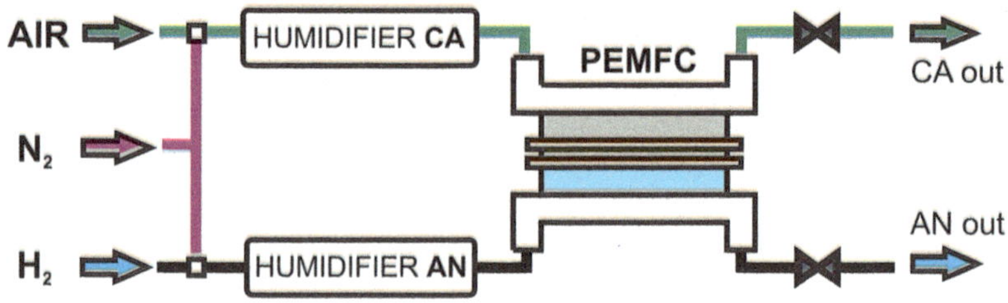

Fig. 7.6 Greenlight Innovation G60 humidification process

Acquisition and automation subsystem enables the end user to successfully run testing scripts and log all data readings from installed sensors. The whole setup is established as a supervisory control and data acquisition system (SCADA). The automation system is realized using the Emerald software package, provided by Greenlight Innovation. Subsystem control loops (closed and open), such as pressure control loop, flow control loop, temperature control loop, and humidity control loop, are completely customizable, and user can freely set up and tune controller parameters for a certain loop. This is very important for the experimental validation of the theoretically obtained models, because experimental setup can be run in a regime of interest in controlled environment. Furthermore, the number of acquisition channels is expandable so that new sensors could be added quite easily.

The safety features are designed to prevent all hazardous situations that might occur during operation. Those features are:

1. Hard wired interlocks.
2. Factory software interlocks.
3. User-configurable software interlocks (warning high/law alarms and shut-down high/low alarms).
4. Nitrogen purge system (nitrogen is used as nonreacting gas), schematically shown in Fig. 7.5.
5. Hydrogen sensor to detect possible leaks.
6. Manual emergency stop button.

From the academic and educational point of view, this station represents a very advanced experimental setup, which can provide precise, valid, and valuable results. Safety features provide high level of safety, and this station can be used comfortably by both undergraduate and graduate students.

Greenlight TP50 Fuel Cell is a robust research and development tool that uses well established materials, and in combination with quick assembly features, it is well suited for sample screening, research and development, and educational purposes. Fuel cell assembly has heated endplates that provide and maintain the stack temperature at the desired level. Also, there are access points on the cathode and anode endplate heaters for thermocouple placement and temperature acquisition. This type of fuel cell can be assembled independently of the compression hardware that makes it very easy to set up, expand, and test various stack configurations.

TP50 is a modular setup which in our case has three individual cells stacked up with a total active area of 50 cm^2. The PEM fuel cell is built with materials which have proven compatibility with fluids used for chemical reaction occurring in it. The sealing gaskets are very reliable and can accommodate a variety of fuel cell membrane types such as Nafion 112, 115, 117, etc. The membrane on our experimental setup is Nafion 212. This setup is very reliable and robust, and in terms of maintenance, the only thing that requires regular replacements is the membrane in order to achieve the best performance. For research and educational purposes, where exploitation is extreme, and tear and wear of the membrane are affecting performance of the fuel cell, this setup is long-term sustainable with modest maintenance costs.

We have developed in (Milanovic et al. 2017) a fifth-order nonlinear mathematical model for the considered fuel cell that agrees very well with experimental results. The model will be presented in the next subsection where we provide also a general review of modeling of PEMFC. The state space variables are $m_{O_2}(t)$, mass of oxygen in cathode; $m_{N_2}(t)$, mass of nitrogen in cathode; $m_{H_2}(t)$, mass of hydrogen in anode; $m_{v,an}(t)$, mass of water vapor in anode; and $m_{v,ca}(t)$, mass of water vapor in cathode. The model has two control inputs: $W_{H_2,in}(t)$, mass flow rate of dry hydrogen supply (fuel), and $W_{a,ca,in}(t)$, mass flow rate of dry air supply (oxygen). The produced fuel cell stack current $I_{st}(t)$ is also an input to the fuel cell. It is treated as a disturbance $w(t)$, that is, $w(t) = I_{st}(t)$, since it changes randomly depending on the fuel cell load. The modeling is done under the ideal gas assumption, a basic assumption used in thermodynamics; Sonntag et al. (1998) states that $pV = nRT$, where p is gas pressure, V represents volume, n stands the number of moles, and T is the constant temperature.

7.3.1 TP50 PEMFC Modeling

In this section, we will describe the modeling steps of a PEMFC and indicate its complexity, by presenting all variables, parameters, and constants used in this process, identifying also the state space variables, fuel cell inputs, fuel cell outputs, and fuel cell disturbance. The modeling follows our recent paper Milanovic et al. (2017).

The mathematical model for the PEMFC in our laboratory is obtained using the basic formulas presented in the fundamental of thermodynamics book (Sonntag et al. 1998), modeling and control book of PEMFC (Purkushpan et al. 2004a), and modeling PEMFC papers (Grujicic et al. 2004a, b; Na et al. 2007; Na and Gou 2008; Kunusch et al. 2011). The derivations of the fifth-order nonlinear mathematical model for G60 PEMFC fuel cell start with the following state space differential equations:

$$\frac{dm_{O_2}(t)}{dt} = W_{O_2,in}(t) - W_{O_2,out}(t) - W_{O_2,rct}(t) \quad (7.51)$$

$$\frac{dm_{N_2}(t)}{dt} = W_{N_2,in}(t) - W_{N_2,out}(t) \quad (7.52)$$

$$\frac{dm_{H_2}(t)}{dt} = W_{H_2,in}(t) - W_{H_2,out}(t) - W_{H_2,rct}(t) \quad (7.53)$$

$$\frac{dm_{v,an}(t)}{dt} = W_{v,an,in}(t) - W_{v,an,out}(t) - W_{v,mem}(t) \quad (7.54)$$

$$\frac{dm_{v,ca}(t)}{dt} = W_{v,ca,in}(t) - W_{v,ca,out}(t) + W_{v,ca,gen}(t) - W_{v,mem}(t) \quad (7.55)$$

where W denotes the mass flow rate and the subscripts *in,out,rct,gen,mem* represent into the cathode/anode, out of the cathode/anode, reacted, generated, and membrane, respectively. The mass flow rate $W_{O_2,rct}$ in (7.51) provides the mass flow rate of oxygen that is lost due to chemical reactions occurring in the fuel cell. $W_{v,ca,gen}$ in (7.55) is the mass flow rate of water vapor gained from the chemical production of water.

Using the Faraday's law of electrolysis, we have

$$W_{O_2,rct} = M_{O_2}\frac{nI_{st}}{4F}, \quad W_{v,ca,gen} = M_v\frac{nI_{st}}{2F} \quad (7.56)$$

where n is the number of individual fuel cells in the stack, I_{st} is the stack current, and F is the Faraday's constant. Mass flow $W_{v,mem}$ in (7.55) represents water vapor membrane transport due to two phenomena, known as electroosmotic drag and water back diffusion. These phenomena are caused by the water vapor presence on both sides of the fuel cell membrane. Detailed derivations for mass flow through membrane can be found in (Grujicic et al. 2004a, b; Purkushpan et al. 2004a) as

$$W_{v,mem} = M_v A_{fc} n\left(n_d\frac{i}{F} - D_w\frac{(c_{v,ca} - c_{v,an})}{t_m}\right) \quad (7.57)$$

where A_{fc} is the fuel cell active area, n_d electroosmotic drag coefficient, i is the stack current density, and D_w is the back-diffusion coefficient. $c_{v,ca}$ and $c_{v,an}$ are water concentration in the cathode and anode, and t_m is the membrane thickness. For all of the flow rates into the cathode, the values can be calculated from a mass fraction of the input mass flow rate of air $W_{a,ca,in}$ using the following set of formulas:

$$\begin{aligned} &W_{O_2,in} = x_{O_2}W_{a,ca,in}, \quad W_{N_2,in} = x_{N_2}W_{a,ca,in} \\ &W_{v,ca,in} = \frac{M_v}{M_a}\frac{p_{v,ca,in}}{(p_{ca,in} - p_{v,ca,in})}W_{a,ca,in} \end{aligned} \quad (7.58)$$

where x_{O_2} and x_{N_2} are mass fractions of O_2 and N_2. M_v and M_a are the molar masses of water vapor and air. $p_{v,ca,in}$ is inlet vapor pressure. Cathode inlet pressure $p_{ca,in}$ is considered to be constant. The mass fractions of gases are defined by

$$x_{O_2} = \frac{y_{O_2} M_{O_2}}{M_a}, x_{N_2} = \frac{(1 - y_{O_2}) M_{N_2}}{M_a} \tag{7.59}$$

where y_{O_2} is the mole fraction of oxygen, which is equal to 0.21 when air is used. Input air, oxygen and nitrogen, has the molar mass M_a as

$$M_a = x_{O_2} M_{O_2} + (1 - x_{O_2}) M_{N_2} \tag{7.60}$$

The partial pressure of the vapor coming into the cathode is given by the following formula:

$$p_{v,ca,in} = \varphi_{ca,in} p_{sat}^{T_{ca,in}} \tag{7.61}$$

where $\varphi_{ca,in}$ is the relative humidity of the incoming gas mixture and $p_{sat}^{T_{ca,in}}$ is the saturation pressure for the gas mixture at the dew point inlet set temperature, $T_{ca,in}$, calculated as (Nguyen and White 1993)

$$\log_{10}(p_{sat}) = -1.69 \times 10^{-10} T^4 + 3.85 \times 10^{-7} T^3 - 3.39 \times 10^{-4} T^2 + 0.143T - 20.92 \tag{7.62}$$

The mass flow rates for O_2, N_2 and water vapor out of the cathode satisfy the following expressions:

$$W_{O_2,out} = \frac{m_{O_2}}{m_{ca}} W_{ca,out}, \quad W_{N_2,out} = \frac{m_{N_2}}{m_{ca}} W_{ca,out}, \quad W_{v,ca,out} = \frac{m_{v,ca}}{m_{ca}} W_{ca,out} \tag{7.63}$$

The total cathode mass mixture m_{ca} is the sum of three state variables and given by the following formula:

$$m_{ca} = m_{O_2} + m_{N_2} + m_{v,ca} \tag{7.64}$$

$W_{ca,out}$ is the total mass flow rate out of the cathode represented by

$$W_{ca,out} = k_{ca}(p_{ca} - p_{ca,out}) \tag{7.65}$$

where k_{ca} is an experimentally derived cathode orifice constant, p_{ca} is the total pressure in the cathode, and $p_{ca,out}$ is the cathode outlet pressure. The total pressure in the cathode p_{ca} can be calculated using the ideal gas law for each gas in the mixture

$$p_i = \frac{m_i R_i T_{st}}{V_{ca}}, \quad i = O_2 \text{ or } N_2, \quad p_{ca} = p_{O_2} + p_{N_2} + p_{v,ca} \tag{7.66}$$

The cathode relative humidity φ_{ca} is assumed to be less than 1, which means that the model assumes that there is no condensation of vapor in the cathode. From the ideal gas law (Sonntag et al. 1998), the relative humidity is

$$p_{v,ca} = \varphi_{ca} p_{sat}^{st}, \quad \varphi_{ca} = \frac{R T_{st} m_{v,ca}}{M_v V_{ca} p_{sat}^{st}} \tag{7.67}$$

where p_{sat}^{st} is calculated from (7.61) at the stack temperature T_{st}.

The *anode side model* assumes that pure hydrogen is used as a fuel so that the hydrogen flow rate into the anode is equal to the input mass flow rate of hydrogen

$$W_{H_2,in} = W_{H_2,an,in} \tag{7.68}$$

The mass flow of vapor into the anode is given by

$$W_{v,an,in} = \frac{M_v}{M_{H_2}} \frac{p_{v,an,in}}{\left(p_{an,in} - p_{v,an,in}\right)} W_{H_2,an,in} \tag{7.69}$$

where M_v and M_{H_2} are molar masses of water vapor and hydrogen and $p_{v,an,in}$ is inlet vapor pressure. Anode inlet pressure $p_{an,in}$ is considered to be constant. The vapor pressure of the gas coming into the anode $p_{v,an,in}$ is given as (Sonntag et al. 1998)

$$p_{v,an,in} = \varphi_{an,in} p_{sat}^{T_{an,in}} \tag{7.70}$$

where $\varphi_{an,in}$ is the relative humidity of the incoming gas mixture and $p_{sat}^{T_{an,in}}$ is the saturation pressure for the gas mixture at the dew point inlet set temperature $T_{an,in}$, calculated from (7.61). The mass flow rates of H_2 and water vapor out of the anode are given by

$$W_{H_2,out} = \frac{m_{H_2}}{m_{an}} W_{an,out}, \quad W_{v,an,out} = \frac{m_{v,an}}{m_{an}} W_{an,out} \tag{7.71}$$

where the total anode mass mixture m_{an} is a sum of the two state variables m_{H_2} and $m_{v,an}$

$$m_{an} = m_{H_2} + m_{v,an} \tag{7.72}$$

In (7.71), $W_{an,out}$ is the total mass flow rate out of the anode and is given by the following formula:

$$W_{an,out} = k_{an}\left(p_{an} - p_{an,out}\right) \tag{7.73}$$

where k_{an} is an experimentally derived anode orifice constant, p_{an} is the total pressure in the anode, and $p_{an,out}$ is the anode outlet pressure. The total pressure in the anode p_{an} can be found using the ideal gas law as

$$p_{H_2} = \frac{m_{H_2} R_{H_2} T_{st}}{V_{an}}, \quad p_{an} = p_{H_2} + p_{v,an} \tag{7.74}$$

The pressure of vapor in the anode $p_{v,an}$ can be found as

$$p_{v,an} = \varphi_{an} p_{sat}^{st} \tag{7.75}$$

The anode relative humidity φ_{an} is assumed to be less than 1, which means that the model assumes that there is no condensation of vapor in the anode. From the ideal gas law, the relative humidity is (Sonntag et al. 1998)

$$\varphi_{an} = \frac{RT_{st} m_{v,an}}{M_v V_{an} p_{sat}^{st}} \tag{7.76}$$

The mass flow rate of hydrogen $W_{H_2,rct}$, present in differential equation (7.53), that is lost in the chemical reaction occurring in the fuel cell can be found from the Faraday's laws as

$$W_{H_2,rct} = M_{H_2} \frac{nI_{st}}{2F} \tag{7.77}$$

where M_{H_2} is the molar mass of hydrogen. Mass flow $W_{v,mem}$ is negative in (7.54) because this flow is from anode to cathode.

In the state space fuel cell model (7.51, 7.52, 7.53, 7.54 and 7.55), the model state, control input, model output, and disturbance (produced fuel cell current since it comes back to the fuel cell input) are, respectively, defined by

$$x(t) = \left[m_{O_2}(t) \quad m_{N_2}(t) \quad m_{H_2}(t) \quad m_{v,an}(t) \quad m_{v,ca}(t) \right]^T \tag{7.78}$$

$$u(t) = \left[W_{H_2,an,in}(t) \quad W_{a,ca,in}(t) \right]^T \tag{7.79}$$

$$y(t) = v_{st}(t) \tag{7.80}$$

$$w(t) = I_{st}(t) \tag{7.81}$$

where $v_{st}(t)$ is the fuel cell stack voltage.

The single-cell voltage can be written as

$$v_{fc} = E - v_{act} - v_{ohm} - v_{conc} \tag{7.82}$$

where E is the open circuit voltage generated in the fuel cell electrochemical reaction. v_{act}, v_{ohm}, and v_{conc} are three voltage losses known as activation loss, ohmic loss, and concentration loss. For n fuel cell plates in a stack, the stack voltage is given by $v_{st} = nv_{fc}$.

The open circuit voltage produced by the fuel cell plate denoted by E is given in terms of the standard Gibbs free energy ΔG^o, temperature T, pressures of hydrogen p_{H_2}, oxygen p_{O_2}, and water vapor p_{H_2O}, via the well-known Nernst formula

$$E = -\frac{\Delta G^o}{2F} + \frac{RT}{2F}\ln\left(\frac{p_{H_2}p_{O_2}^{0.5}}{p_{H_2O}}\right) \tag{7.83}$$

This formula assumes a reversible process system. However, in reality there are energy losses such as energy conversion into heat. It also assumes standard conditions for ΔG^o, when fuel cell operating conditions may be different. Formula (7.83) can be expanded to factor in these energy losses and operating conditions, Pukrushpan et al. (2004a).

Mathematical formulas for the considered PEM fuel cell activation voltage loss, ohmic voltage loss, and concentration voltage loss that appear in formula (7.82) are rather lengthy. They can be found in (Milanovic et al. 2017).

7.3.2 *Simulation Results*

The nonlinear system model is linearized around the steady operating point 0.5 A/cm^2 (stack current: 25 A). This operating point is in ohmic polarization region for the fuel cell, Milanovic et al. (2017), which is a desirable operating regime for PEM fuel cells, Benziger et al. (2006). The steady-state operating point is taken as (Milanovic and Radisavljevic-Gajic 2018)

$$\begin{bmatrix} m_{O_2}^{ss} \\ m_{N_2}^{ss} \\ m_{H_2}^{ss} \\ m_{v,an}^{ss} \\ m_{v,ca}^{ss} \end{bmatrix} = \begin{bmatrix} 1.518 \times 10^{-6} \\ 9.919 \times 10^{-6} \\ 8.558 \times 10^{-7} \\ 7.566 \times 10^{-7} \\ 2.003 \times 10^{-6} \end{bmatrix} [\mathrm{kg}]$$

$$\begin{bmatrix} u_1^{ss} \\ u_2^{ss} \end{bmatrix} = \begin{bmatrix} 2.49 \\ 0.78 \end{bmatrix} [\mathrm{NLPM}], \quad v_{ss} = 1.875[\mathrm{V}], \quad I_{st}^{ss} = 25\,[\mathrm{A}]$$

The linearized model state space matrices are given by

$$A = \begin{bmatrix} -110.936 & -122.752 & 0 & 0 & -190.96 \\ -701.503 & -805.673 & 0 & 0 & -1247.78 \\ 0 & 0 & -1.2211 & 0.1905 & 0 \\ 0 & 0 & -0.5855 & -151.17 & 56.659 \\ 141.658 & -161.972 & 0 & 150.58 & -312.21 \end{bmatrix}$$

$$C = [\,191912.69 \quad -6184.19 \quad 53377.42 \quad -2299.14 \quad -10567.25\,], \quad D = \begin{bmatrix} 0 \\ 0 \end{bmatrix}$$

$$B = \begin{bmatrix} -3.1277 & 0 \\ -21.1894 & 0 \\ 0 & 0.9838 \\ 0 & 0.3272 \\ -4.4095 & 0 \end{bmatrix}$$

The developed mathematical model for the PEM fuel cell was experimentally validated in our fuel cell laboratory. The parameters and constants used in the experiment are presented in Tables 7.2 and 7.3.

We have performed simulation for the considered model using the software package MATLAB/Simulink and compared obtained results with the actual experimental results. It can be seen from Fig. 7.7 that the model output matches very well with the experimental data output. MATLAB® System Identification Toolbox also

Table 7.2 PEM fuel cell modeling general parameters

Parameter	Symbol	Value	SI units
Atmospheric pressure	p_{atm}	101,325	Pa
Atmospheric temperature	T_{atm}	298.15	K
Specific heat ratio of air	γ	1.4	–
Specific heat of air	C_p	1004	J/kg/K
Density of air	ρ_a	1.23	kg/m^3
Universal gas constant	R	8.314	J/mol/K
Air gas constant	R_a	286.9	J/kg/K
Oxygen gas constant	R_{O_2}	259.8	J/kg/K
Nitrogen gas constant	R_{N_2}	296.8	J/kg/K
Hydrogen gas constant	R_{H_2}	4124.3	J/kg/K
Water vapor gas constant	R_v	461.5	J/kg/K
Molar mass of air	M_a	28.97×10^{-3}	kg/mol
Molar mass of oxygen	M_{O_2}	32.0×10^{-3}	kg/mol
Molar mass of nitrogen	M_{N_2}	28.0×10^{-3}	kg/mol
Molar mass of hydrogen	M_{H_2}	2.0×10^{-3}	kg/mol
Molar mass of water vapor	M_v	18.02×10^{-3}	kg/mol
Faraday's constant	F	96,487	A·s/mol

Table 7.3 PEM fuel cell modeling system parameters

Parameter	Symbol	Value	SI units
Number of cells in fuel cell stack	n	3	–
Temperature of fuel cell stack	T_{st}	353	K
Fuel cell active area	A_{fc}	50	cm^2
Stack cathode volume	V_{ca}	8.1419×10^{-6}	m^3
Stack anode volume	V_{an}	3.3928×10^{-6}	m^3
Anode outlet orifice constant	k_{an}	0.4177×10^{-1}	kg/s/Pa
Cathode outlet orifice constant	k_{ca}	0.8470×10^{-7}	kg/s/Pa
Membrane dry density	$\rho_{mem,dry}$	0.002	kg/cm^3
Membrane dry equivalent weight	$M_{mem,dry}$	1.1	kg/mol
Membrane thickness	t_m	1.275×10^{-2}	cm
Average ambient air relative humidity	φ_{atm}	0.5	–
Oxygen mole fraction at cathode inlet	y_{O_2}	0.21	–

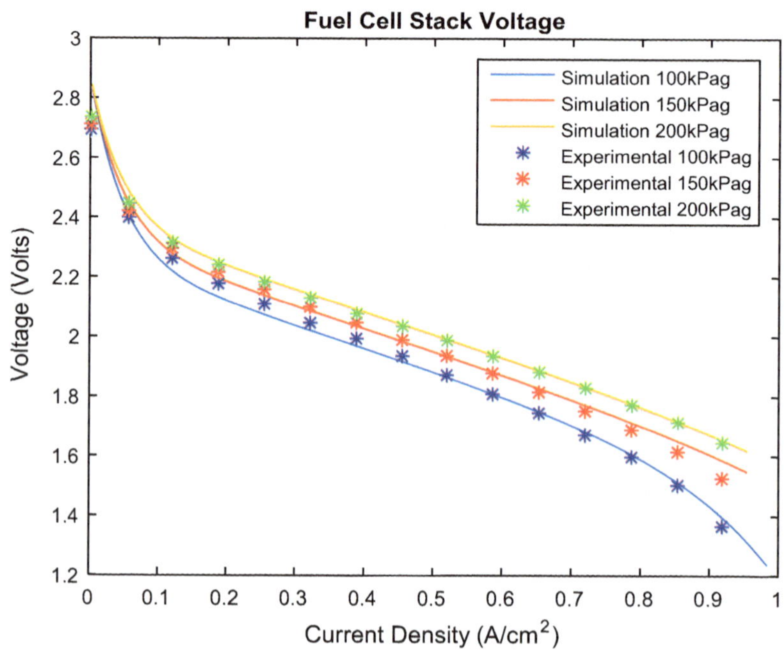

Fig. 7.7 Simulation and experimental curves for various inlet pressures for a three-plate stack

produced percentage of the model fit with experimental data. The model has 90% fit for 100 kPag inlet pressures and 92% fit for 200 kPag with the experimental data. It can be seen that the model is most accurate within the current density range of 0.5 A/cm^2 to $0.8 A/cm^2$, which is the typical range of operation for fuel cell testing.

The developed model is control oriented, which means that it can be further studied from the control systems point of view. Practical implementation issues on the modeled system include flooding, gas starvation, overheating, membrane humidity (concentration of water vapor on the anode and cathode sides), etc. Those issues affect the performance and longevity of the fuel cell stack. To overcome these obstacles, adequate control strategies should be used to efficiently operate the fuel cell system.

Control approaches to PEM fuel cells that can be found in the literature are summarized in the recent paper (Daud et al. 2017). They could be divided into the following categories:

- Classical approaches, feedback and feedforward controller designs, and implementation.
- Adaptive control approach.
- Model predictive control approach.
- Neural network and fuzzy logic approach.
- Sliding mode control.

7.4 A Fifth-Order Nonlinear PEMFC Model

Another fifth-order nonlinear fuel cell model was considered by Na and Gou (2008) and Gao et al. (2010). The model basically has the same starting differential equations for the five state variables as in Sect. 7.3, with gas masses replacing gas pressures. Due to a simple relation that exists between the gas mass and gas pressure (under constant temperature)

$$m(t) = p(t)\frac{V}{RT} \tag{7.84}$$

differential equations (7.51, 7.52, 7.53, 7.54 and 7.55) can be expressed in terms of gas pressures, which leads to the following mathematical models:

$$\frac{dp_{\mathrm{H}_2}(t)}{dt} = \frac{RT}{V_A}(W_{\mathrm{H}_2,in}(t) - W_{\mathrm{H}_2,out}(t) - W_{\mathrm{H}_2,rct}(t)) \tag{7.85}$$

$$\frac{dp_{v,an}(t)}{dt} = \frac{RT}{V_A}(W_{v,an,in}(t) - W_{v,an,out}(t) - W_{v,mem}(t)) \tag{7.86}$$

$$\frac{dp_{\mathrm{O}_2}(t)}{dt} = \frac{RT}{V_C}(W_{\mathrm{O}_2,in}(t) - W_{\mathrm{O}_2,out}(t) - W_{\mathrm{O}_2,rct}(t)) \tag{7.87}$$

$$\frac{dp_{N_2}(t)}{dt} = \frac{RT}{V_C}(W_{N_2,in}(t) - W_{N_2,out}(t)) \tag{7.88}$$

$$\frac{dp_{v,ca}(t)}{dt} = \frac{RT}{V_C}\left(W_{v,ca,in}(t) - W_{v,ca,out}(t) + W_{v,ca,gen}(t) - W_{v,mem}(t)\right) \tag{7.89}$$

Note different ordering of state space variables in (7.51, 7.52, 7.53, 7.54 and 7.55) and (7.85, 7.86, 7.87, 7.88 and 7.89), which is consistent with the ordering of the state space variables in the model derived by Na and Gou (2008).

The model of Na and Gou (2008) assumes that

$$W_{v,mem} = 1.2684\frac{N}{F}I \approx 5K_rI, \quad K_r = \frac{N}{4F} \tag{7.90}$$

where N stands for the number of fuel cells in the stock. In addition, it is well-known that the following expressions hold (Larminie and Dicks 2001; Barbir 2005):

$$W_{H_2,rct} = \frac{N}{2F} = 2K_rI, \quad W_{H_2O,rct} = \frac{N}{2F} = 2K_rI, \quad W_{O_2,rct} = \frac{N}{4F} = K_rI \tag{7.91}$$

The outlet molar flow rates for the anode side are given by

$$\begin{aligned} W_{H_2,out} &= (\text{Anode}_{in} - W_{H_2,rct}) \times F_{H_2} \\ W_{v,an,out} &= (\text{Anode}_{in} - W_{v,an,rct}) \times F_{H_2O_A} \end{aligned} \tag{7.92}$$

and for the cathode side, they are given by

$$\begin{aligned} W_{v,ca,out} &= (\text{Cathode}_{in} - W_{v,ca,rct}) \times F_{H_2O_C} \\ W_{O_2,out} &= (\text{Cathode}_{in} - W_{O_2,rct}) \times F_{O_2} \end{aligned} \tag{7.93}$$

with the nitrogen outlet molar flow rate equal to

$$W_{N_2,out} = \text{Cathode}_{in} \times F_{N_2} \tag{7.94}$$

In formulas (7.92, 7.93 and 7.94), F_{i_j} are the pressure fraction factors. Those fractions were introduced for the first time in the fuel cell modeling work of Chiu et al. (2004). In the model considered in this section, they are given by

$$\begin{aligned} F_{H_2} &= \frac{p_{H_2}}{p_{H_2} + p_{H_2O_A}}, \quad F_{H_2O_A} = \frac{p_{H_2O_A}}{p_{H_2} + p_{H_2O_A}} \\ F_{O_2} &= \frac{p_{O_2}}{p_{O_2} + p_{H_2O_C} + p_{N_2}}, \quad F_{N_2} = \frac{p_{N_2}}{p_{O_2} + p_{H_2O_C} + p_{N_2}} \\ F_{H_2O_C} &= \frac{p_{H_2O_C}}{p_{O_2} + p_{H_2O_C} + p_{N_2}} \end{aligned} \tag{7.95}$$

Introducing the notation for the state space variables as

$$\begin{aligned} x(t) &= \left[p_{\mathrm{H}_2}(t) \quad p_{v,an}(t) \quad p_{\mathrm{O}_2}(t) \quad p_{\mathrm{N}_2}(t) \quad p_{v,ca}(t)\right]^T \\ &= \left[x_1(t) \quad x_2(t) \quad x_3(t) \quad x_4(t) \quad x_5(t)\right]^T \end{aligned} \tag{7.96}$$

five nonlinear state space differential equations were derived by Na and Gou (2008)

$$\frac{dx_1(t)}{dt} = \frac{RT}{V_A}\left(W_{\mathrm{H}_2,in}(t) - (\text{Anode}_{in} - 2K_rI)\frac{x_1(t)}{x_1(t)+x_2(t)} - 2K_rI\right) \tag{7.97}$$

$$\frac{dx_2(t)}{dt} = \frac{RT}{V_A}\left(W_{v,an,in}(t) - (\text{Anode}_{in} - 5K_rI)\frac{x_1(t)}{x_1(t)+x_2(t)} - 5K_rI\right) \tag{7.98}$$

$$\frac{dx_3(t)}{dt} = \frac{RT}{V_C}\left(W_{\mathrm{O}_2,in}(t) - (\text{Cathode}_{in} - K_rI)\frac{x_3(t)}{x_3(t)+x_4(t)+x_5(t)} - K_rI\right) \tag{7.99}$$

$$\frac{dx_4(t)}{dt} = \frac{RT}{V_C}\left(W_{\mathrm{N}_2,in}(t) - (\text{Cathode}_{in})\frac{x_4(t)}{x_3(t)+x_4(t)+x_5(t)}\right) \tag{7.100}$$

$$\frac{dx_5(t)}{dt} = \frac{RT}{V_C}\left(W_{v,ca,in}(t) - (\text{Cathode}_{in} + 2K_rI)\frac{x_5(t)}{x_3(t)+x_4(t)+x_5(t)} + 2K_rI - 5K_rI\right) \tag{7.101}$$

The inlet flow rates are given in terms of mole fractions

$$W_{\mathrm{H}_2,in} = 0.99 \times \text{Anode}_{in}, W_{\mathrm{O}_2,in} = 0.21 \times \text{Anode}_{in}, W_{\mathrm{N}_2,in} = 0.79 \times \text{Anode}_{in} \tag{7.102}$$

The control variables were introduced by Na and Gou (2008) in the nonlinear model defined in (7.97, 7.98, 7.99, 7.100, 7.101 and 7.102) as

$$u_a(t) = \frac{1}{k_a}\text{Anode}_{in}, \quad u_c(t) = \frac{1}{k_c}\text{Cathode}_{in} \tag{7.103}$$

where k_a and k_c are appropriate constants.

The inlet water flow rates are given in terms of humidity of anode and cathode φ_A, φ_C,, saturation pressure P_{sat}, and total pressures in anode and cathode, P_A, P_C, as follows:

$$\begin{aligned} W_{v,an,in}(t) &= \frac{\varphi_A P_{sat}}{P_A(t) - \varphi_A P_{sat}}\text{Anode}_{in}(t) \\ W_{v,ca,in}(t) &= \frac{\varphi_C P_{sat}}{P_C(t) - \varphi_C P_{sat}}\text{Cathode}_{in}(t) \end{aligned} \tag{7.104}$$

The total anode and cathode pressures are defined by

$$\begin{aligned} P_A(t) &= p_{H_2}(t) + p_{H_2O_A}(t) = x_1(t) + x_2(t) \\ P_C(t) &= p_{O_2}(t) + p_{H_2O_C}(t) + p_{N_2}(t) = x_3(t) + x_4(t) + x_5(t) \end{aligned} \tag{7.105}$$

Using the expressions (7.102, 7.103, 7.104 and 7.105), in differential equations defined in Equations (7.97, 7.98, 7.99, 7.100 and 7.101), the following fifth-order nonlinear state space fuel cell model is obtained:

$$\frac{dx_1(t)}{dt} = \frac{RT}{V_A}\left(0.99k_a u_a(t) - (k_a u_a(t) - 2K_r I)\frac{x_1(t)}{x_1(t) + x_2(t)} - 2K_r I\right) \tag{7.106}$$

$$\begin{aligned} \frac{dx_2(t)}{dt} &= \frac{RT}{V_A} \\ &\times\left(\frac{\varphi_A P_{sat}}{x_1(t) + x_2(t) - \varphi_A P_{sat}} k_a u_a(t) - (k_a u_a(t) - 5K_r I)\frac{x_1(t)}{x_1(t) + x_2(t)} - 5K_r I\right) \end{aligned} \tag{7.107}$$

$$\frac{dx_3(t)}{dt} = \frac{RT}{V_C}\left(0.21k_c u_c(t) - (k_c u_c(t) - K_r I)\frac{x_3(t)}{x_3(t) + x_4(t) + x_5(t)} - K_r I\right) \tag{7.108}$$

$$\frac{dx_4(t)}{dt} = \frac{RT}{V_C}\left(0.79k_c u_c(t) - k_c u_c(t)\frac{x_4(t)}{x_3(t) + x_4(t) + x_5(t)}\right) \tag{7.109}$$

$$\begin{aligned} \frac{dx_5(t)}{dt} &= \frac{RT}{V_C}\frac{\varphi_C P_{sat}}{x_3(t) + x_4(t) + x_5(t) - \varphi_C P_{sat}} k_c u_c(t) \\ &- \frac{RT}{V_C}\left((k_c u_c(t) + 2K_r I)\frac{x_5(t)}{x_3(t) + x_4(t) + x_5(t)} - 3K_r I\right) \end{aligned} \tag{7.110}$$

Using numerical data from Na and Gou (2008), the model (7.106, 7.107, 7.108, 7.109 and 7.110) was linearized in Park and Gajic (2012) at its steady-state points obtained via the MATLAB package for symbolic computations. The steady-state points obtained in Park and Gajic (2012) were given by

$$x^{ss} = \begin{bmatrix} x_1^{ss} & x_2^{ss} & x_3^{ss} & x_4^{ss} & x_5^{ss} \end{bmatrix}^T = \begin{bmatrix} 2.6509 & 0.003 & 7.009 & 26.175 & 0.3390 \end{bmatrix}^T$$

The corresponding linearized system and its state space matrices were obtained as follows:

$$\begin{aligned} \frac{d\Delta x(t)}{dt} &= A\Delta x(t) + B\Delta u(t) + G\Delta I(t) \\ \Delta y(t) &= C\Delta x(t) \end{aligned} \tag{7.111}$$

$$A = \begin{bmatrix} -8.741 \times 10^{-7} & 0.00782 & 0 & 0 & 0 \\ -0.00102 & -0.00884 & 0 & 0 & 0 \\ 0 & 0 & -0.02767 & 0.00732 & 0.00732 \\ 0 & 0 & 0.02733 & -0.00767 & 0.02733 \\ 0 & 0 & 0.00005 & 0.00005 & -0.03508 \end{bmatrix}$$

$$B = \begin{bmatrix} 9.5846 \times 10^{-8} & 0 \\ 8.5750 \times 10^{-4} & 0 \\ 0 & 2.0935 \times 10^{-8} \\ 0 & 2.9127 \times 10^{-8} \\ 0 & -6.3572 \times 10^{-8} \end{bmatrix}, \quad G = \begin{bmatrix} 2.7255 \times 10^{-9} \\ -2.4384 \times 10^{-8} \\ 9.6493 \times 10^{-8} \\ 0 \\ 8.1233 \times 10^{-8} \end{bmatrix}$$

$$C = \begin{bmatrix} 1 & 0 & 0 & 0 & 0 \\ 0 & 0 & 1 & 0 & 0 \end{bmatrix}$$

It is interesting to observe that the linearized system represents a decoupled system, one linear independent subsystem of order two corresponding to the anode dynamics, and another linear independent subsystem of order three corresponding to the cathode dynamics. Hence, controllers can be independently designed for the anode and cathode subsystems of this PEM fuel cell linearized model. This implies the conclusion that the nonlinear fifth-order fuel cell dynamic system is a weakly coupled system (Gajic et al. 2009) having weakly coupled anode and cathode dynamics.

For this *linearized* PEM fuel cell model (7.111), Park and Gajic (2012) developed a sliding mode controller to keep the hydrogen pressure and oxygen pressure at the same value (3 atm) despite piecewise changes in the current (which in fact represents a disturbance to the fuel cell). By keeping the same gas pressures on the both sides of the membrane, the membrane life is extended (Li et al. 2015a, b; Hayati et al. 2016). The design of a sliding mode controller for the fifth-order *nonlinear* fuel cell model (7.106, 7.107, 7.108, 7.109 and 7.110) that keeps efficiently the hydrogen and oxygen pressures at 3 atm despite frequent current (disturbance) changes as a piecewise constant was demonstrated by Park and Gajic (2014).

7.5 Eight-Order Mathematical Model of a PEMFC Used in Electric Cars

In addition to the PEM fuel cell mathematical model developed in Sects. 7.3 and 7.4 that has five state space variables, mass of oxygen in cathode, mass of nitrogen in cathode, mass of hydrogen in anode, mass of water vapor in anode, and mass of water vapor in cathode, the considered eight-order model takes into account its specific automotive application and provides information about dynamics for intake (supply) and outtake (return) manifolds and the compressor used to pump the hydrogen gas. This model was developed by Pukrushpan et al. (2004a, b) and used in many journal papers dealing with modeling and control of PEM fuel cells.

The supply manifold is modeled by differential equations that provide information about evolution of the gas mass $m_{sm}(t)$ and gas pressure $p_{sm}(t)$ in the supply manifold. Note that due to temperature changes in the supply manifold, the relation among these two state space variables is not any more the simple one obtained from the ideal gas law. The return manifold has constant temperature so that only one scalar differential equation is needed to completely model its dynamics for the gas pressure $p_{rm}(t)$. The compressor dynamics is also modeled by one scalar differential equation representing the dynamics of the blower angular speed $\omega_{cp}(t)$. For the corresponding differential equations for $m_{sm}(t)$, $p_{sm}(t)$, $p_{rm}(t)$, and $\omega_{cp}(t)$, the reader is referred to (Grujicic et al. 2004a, b; Pukrushpan et al. 2004a, b).

Note that the overall model is nonlinear and of order nine. It was shown by Pukrushpan et al. (2004a, b) that this model can be linearized around its nominal operating points and even its dynamics represented very well by an eighth-order linear mathematical model. Namely, dynamics of water mass in the cathode is negligible since it is both weakly controllable and weakly observable (Zhou and Doyle 1998). More about the original ninth-order nonlinear model of Pukrushpan et al. (2004a, b) including all state space variables, and its linearized eight-order mathematical model, including the linearized model state space matrices, is presented in Sect. 2.4.1 of this monograph.

7.6 Notes

Presentation of material in Sect. 7.1 is based on the journal paper by Radisavljevic (2011). Presentations in Sects. 7.2 and 7.3 follow the conference papers of Radisavljevic-Gajic and Graham (2017) and Milanovic et al. (2017). Permissions for the use of such material in this research monograph were granted to us by Elsevier for the paper by Radisavljevic (2011) published in *Journal of Power Sources* and by the American Society of Mechanical Engineers (ASME) for two conference papers by Radisavljevic-Gajic and Graham (2017) and Milanovic et al. (2017) presented at the 2017 ASME Dynamic Systems and Control Conference.

Chapter 8
Control of a Hydrogen Gas Processing System

In this chapter, we present a reduced-order observer-driven controller design for a linearized model of a fuel cell hydrogen gas processing system (also called a hydrogen gas reformer or simply a reformer), which produces hydrogen from natural gas. To solve this control problem, we first design a reduced-order observer to estimate state space variables needed for feedback at all times. Then, we design two feedback control loops, one of them with an integrator (integral control, Khalil 2002; Sinha 2007) and another one with proportional feedback from the estimated state variables (obtained from the observer). In the third step, a feed-forward controller is designed whose role is to offset for the impact of the disturbance caused by the fuel cell current. Both the feedback controller and the feed-forward controller are obtained through a rigorous dynamic optimization process of a quadratic performance criterion along trajectories of a linear continuous-time dynamic system. According to the presented simulation results, the proposed controller clearly copes well with the disturbance and reduces its impact within a few seconds from the time when the disturbance occurs, despite large jumps in the fuel cell current (disturbance). Even more, it outperforms the corresponding full-order observer-based controller developed in Pukrushpan et al. (2004a, b) for the same hydrogen gas reformer processing system. The use of observers in controlling fuel cells was considered also in Pilloni et al. (2015).

The presentation of this chapter mostly follows the author's recent paper, Radisavljevic-Gajic and Rose (2015), with some additional explanations and clarifications.

V. Radisavljević-Gajić et al., *Multi-Stage and Multi-Time Scale Feedback Control of Linear Systems with Applications to Fuel Cells*, Mechanical Engineering Series,
https://doi.org/10.1007/978-3-030-10389-7_8

8.1 Introduction

A PEM fuel cell hydrogen gas reformer that produces hydrogen used in PEMFC from natural gas has been considered in several papers (see, e.g., Pukrushpan et al. 2004a, b, 2006; Tsourapas et al. 2007; Cipiti et al. 2013). The hydrogen obtained is pumped to the anode side of the PEM fuel cell.

Importance of fuel cells as green electric energy generators has been nicely demonstrated in the book of Hoffmann and Dorgan (2012). PEM fuel cells are the most developed and the best understood among all fuel cells, and they can be used for both mobile (vehicles, portable computing devices) and stationary applications (residential and industrial electric power generation, data centers). It is interesting to point out that in December of 2011, Apple Inc. filled a patent on portable computing devices, Spare et al. (2011), which will use fuel cells. The efficient use of renewable energy for Internet Protocol (IP) over wavelength division multiplexing (WDM) fiber-optic network for data centers was demonstrated in Dong et al. (2012).

In addition to the hydrogen gas reformer whose mathematical model was considered in Pukrushpan et al. (2004a, b, 2006), Tsourapas et al. (2007), and Cipiti et al. (2013), which will be the main subject of this chapter, the study of hydrogen gas reformers (also known as fuel processors or fuel processing systems) for PEM fuel cells, both experimentally and via development of corresponding mathematical models, has been considered in several additional papers (Zhu et al. 2001; Seo et al. 2006; Mitchell et al. 2006; Lee et al. 2007; Adachi et al. 2009; Karstedt et al. 2011; Sciazko et al. 2014), including those for vehicular applications (zur Megede 2002; Severin et al. 2005).

Fuel cells utilize chemical reactions with hydrogen gas to produce electricity. However, H_2 gas is not always easily available for a fuel cell system. A solution to this problem is to use a hydrogen gas reformer, also known as the fuel processor system (FPS) to purify gas, natural gas, into the needed H_2 gas (Pukrushpan et al. 2004a, b, 2006; Tsourapas et al. 2007; Cipiti et al. 2013). A common process used in a FPS is partial oxidation. This process uses chemical reactions of natural gas and air to produce a H_2-rich gas product. The four main reactors of the FPS presented in Sect. 2.3.1 in Fig. 2.1 are hydro-desulfurizer (HDS), catalytic partial oxidizer (CPOX), water gas shift (WGS), and preferential oxidizer (PROX). Once the gas travels through all of these reactors, H_2-rich gas will be produced.

Natural gas enters the FPS via a high pressure source, usually a tank or a gas line. The gas is first fed through the hydro-desulfurizer to eliminate any sulfur that could be contained in the gas. This is done because sulfur can poison the water gas shift. The desulfurized gas is then brought to the mixer, where it blends with air. The air is first brought into the FPS by a blower and then passes through a heat exchanger (HEX) to reach a necessary temperature.

Once mixed, the gas passes through the catalytic partial oxidizer where a catalyst causes the natural gas to react with the oxygen in the air. Two exothermal reactions take place in

$$\text{(CPOX)}: \text{ Partial oxidation (POX) and total oxidation (TOX)}$$

Partial oxidation produces H_2 gas and carbon monoxide. The total oxidation produces water and carbon dioxide. Even though both reactions generate heat, TOX releases much larger amount of heat ($\Delta H^o_{tox} = -0.8026 \times 10^6\ \text{J/mol}$).

$$\text{(POX)} \qquad CH_4 + \frac{1}{2}O_2 \rightarrow CO + 2H_2 \qquad \Delta H^o_{pox} = -0.036 \times 10^6\ \text{J/mol}$$

$$\text{(TOX)} \qquad CH_4 + 2O_2 \rightarrow CO_2 + 2H_2O \qquad \Delta H^o_{tox} = -0.8026 \times 10^6\ \text{J/mol}$$

Since only POX is producing H_2, it is preferable to increase the amount of gas reacting through POX instead of TOX. That is highly dependent on the ration of O_2 and gas entering CPOX and the CPOX catalyst bed temperature T_{cpox}, Zhu et al. (2001). In addition, very high-temperature T_{cpox} can cause CPOX damage, while the low-temperature T_{cpox} causes inefficient reactions. Since the reactions in the CPOX, especially H_2 production, are strongly dependent on the CPOX reactor temperature T_{cpox}, it is necessary to regulate the temperature efficiently via outside feedback control loops.

Even though H_2 is produced by the POX reaction, carbon monoxide is also produced. CO poisons the PEM fuel cell catalyst and therefore needs to be removed. This issue gives need for the next two reactors, the water gas shift and the preferential oxidizer. From the CPOX, the gas mixture flows into the water gas shift, where water is injected into the chamber to react with CO

$$\text{(WGS)} \qquad CO + H_2O \rightarrow CO_2 + H_2$$

The WGS reaction eliminates CO and produces additional H_2. This process does not convert all CO into CO_2, and the mixture is not safe for fuel cell applications. Therefore, the gas mixture is next passed into the PROX, where the remaining CO reacts with the oxygen from the injected air.

$$\text{(PROX)} \quad 2CO + O_2 \rightarrow 2CO_2$$

After leaving the PROX, the gas is rich in H_2 and is safe to be sent to the anode of the PEM fuel cell.

The reformer mathematical model and its controller/observer design techniques are considered in Pukrushpan et al. (2004a, b, 2006), Tsourapas et al. (2007), using the full-order observer. In this chapter, we will show that the controller based on a reduced-order observer produces better results. In addition, we will complement the results of Pukrushpan et al. (2004a, b, 2006), Tsourapas et al. (2007) by formulating more precisely the optimal control problem with integral action and provide the corresponding optimal control strategy, which is used to reject the constant disturbance caused by the produced fuel cell current. The controller designed helps to regulate temperature of the catalytic partial oxidation process and the anode hydrogen mole fraction at the desired values.

The tenth-order nonlinear mathematical model of the considered hydrogen gas reformer was developed in Pukrushpan et al. (2004a, b, 2006) and Tsourapas et al. (2007). Its state space variables are

$$\begin{aligned}\frac{dx(t)}{dt} &= f(x(t), u(t), w(t)) \\ x(t) &= [x_1(t)\ \ x_2(t)\ \ x_3(t)\ \ x_4(t)\ \ x_5(t)\ \ x_6(t)\ \ x_7(t)\ \ x_8(t)\ \ x_9(t)\ x_{10}(t)\,]^T \\ &= \left[\, T_{cpox}\ \ p_{H_2}^{an}\ \ p^{an}\ \ p^{hex}\ \ \omega^{bl}\ \ p^{hds}\ \ p_{CH_4}^{mix}\ \ p_{air}^{mix}\ \ p_{H_2}^{wrox}\ \ p^{wrox} \,\right]\end{aligned} \tag{8.1}$$

The state variables represent the following quantities:

$x_1(t) = T_{cpox}(t)$ – catalyst temperature
$x_2(t) = p_{H_2}^{an}(t)$ – hydrogen pressure in the anode
$x_3(t) = p^{an}(t)$ – anode pressure
$x_4(t) = p^{hex}(t)$ – heat exchange pressure
$x_5(t) = \omega^{bl}(t)$ – compressor blower angular velocity (rad/s)
$x_6(t) = p^{hds}(t)$ – hydro-desulfurizer pressure
$x_7(t) = p_{CH_4}^{mix}(t)$ – pressure in the mixer
$x_8(t) = p_{air}^{mix}(t)$ – air pressure in the mixer
$x_9(t) = p_{H_2}^{wrox}(t)$ – hydrogen pressure in the gas shift converter
$x_{10}(t) = p^{wrox}(t)$ – the total pressure in the gas shift converter

The compressor blows air needed for the fuel (natural gas) oxidation. In model (8.1), $w(t)$ is the disturbance, and it represents the fuel cell stack (connected to the hydrogen reformer) current, $w(t) = I_{st}(t) = V_{st}(t)/R_L$, where $V_{st}(t)$ is the fuel cell stack voltage and R_L stands for the fuel cell stack load.

The control variables operate the blower's angular velocity and the fuel (natural gas) tank valve, that is

$$u(t) = \begin{bmatrix} u_{blower}(t) \\ u_{valve}(t) \end{bmatrix} \tag{8.2}$$

The measured output $y(t)$ serves at the same time as the controlled variable $y(t) = z(t)$. It is defined by

$$y(t) = z(t) = \left[\, T_{cpox}(t)\ \ \ y_{H_2}^{an}(t) \,\right]^T \tag{8.3}$$

where $y_{H_2}^{an}$ is the anode hydrogen mole fraction.

The control goal is to regulate the catalytic partial oxidation temperature and the anode hydrogen mole fraction at the desired values at steady state to $T_{cpox} = 972$ K (that corresponds to the ratio of the number of oxygen moles over the number of methane moles equal to 0.6) and $y_{H_2}^{an} = 0.088$ (8.8%) (corresponding to utilization of

80%). In this section, we consider the hydrogen gas reformer linearized mathematical model, Pukrushpan et al. (2004a, b, 2006), Tsourapas et al. (2007). The linearized system is represented by

$$\begin{aligned}
&\frac{\delta x(t)}{dt} = A\delta x(t) + B\delta u(t) + G\delta w(t)\\
&\delta y(t) = \delta z(t) = C\delta x(t)\\
&y(t) = y_{ss} + \delta y(t), \quad z(t) = z_{ss} + \delta z(t), \quad x(t) = x_{ss} + \delta x(t)\\
&u(t) = u_{ss} + \delta u(t), \quad w(t) = w_{ss} + \delta w(t)
\end{aligned} \tag{8.4}$$

The linearized system matrices are obtained by following the linearization procedure of nonlinear systems at nominal points (see, e.g., Gajic and Lelic 1996; Khalil 2002; Gajic 2003). The linearization is performed at the desired steady-state points whose values can be found in page 147 of Pukrushpan et al. (2004a). The state space matrices obtained are given by

$$A = \begin{bmatrix}
-0.074 & 0 & 0 & 0 & 0 & 0 & -3.53 & 1.0748 & 0 & 0\\
0 & -1.468 & -25.3 & 0 & 0 & 0 & 0 & 0 & 2.5582 & 13.911\\
0 & 0 & -156 & 0 & 0 & 0 & 0 & 0 & 0 & 33.586\\
0 & 0 & 0 & 124.5 & 212.63 & 0 & 112.69 & 112.69 & 0 & 0\\
0 & 0 & 0 & 0 & -3.3333 & 0 & 0 & 0 & 0 & 0\\
0 & 0 & 0 & 0 & 0 & -32.43 & 32.304 & 32.304 & 0 & 0\\
0 & 0 & 0 & 0 & 0 & 331.8 & -344 & -341 & 0 & 9.9042\\
0 & 0 & 0 & 221.97 & 0 & 0 & -253.2 & -254.9 & 0 & 32.526\\
0 & 0 & 2.0354 & 0 & 0 & 0 & 1.8309 & 1.214 & -0.358 & -3.304\\
0.0188 & 0 & 8.1642 & 0 & 0 & 0 & 5.6043 & 5.3994 & 0 & -13.61
\end{bmatrix}$$

$$B = \begin{bmatrix}
0 & 0 & 0 & 0 & 0.12 & 0 & 0 & 0 & 0 & 0\\
0 & 0 & 0 & 0 & 0 & 0.1834 & 0 & 0 & 0 & 0
\end{bmatrix}^T$$

$$G = \begin{bmatrix} 0 & -0.328 & -0.024 & 0 & 0.0265 & 0.0504 & 0 & 0 & 0 & 0 \end{bmatrix}^T$$

$$C = \begin{bmatrix}
1 & 0 & 0 & 0 & 0 & 0 & 0 & 0 & 0 & 0\\
0 & 0.994 & -0.088 & 0 & 0 & 0 & 0 & 0 & 0 & 0
\end{bmatrix}$$

Assuming that the stack current changes as a piecewise constant, that is, $I_{st} = v_{st}/R_L$ (where R_L is the load that changes at some random time instants when users turn on or off), Pukrushpan et al. (2006), the design of controllers with integral action Khalil (2002), Sinha (2007) will be required to cope with the piecewise continuous disturbance $\delta w(t) = \delta I_{st}(t)$. Moreover, for feedback control we will need to estimate all ten state variables based on information coming from the two-dimensional system output. For that purpose we will design a reduced-order observer of dimension eight since the remaining two state variables can be obtained from the output equation. For both the feedback controller design and the reduced-order observer design, we will need that the considered hydrogen gas reformer is both controllable and observable (Chen 2012). The importance of system controllability for fuel cells has been considered in several papers (see, e.g., Serra et al. 2005; McCain et al. 2010; Radisavljevic 2011). The controllability and observability conditions have been verified using MATLAB and the corresponding controllability and observability tests.

Comment 8.1: Controllability and Observability Tests via MATLAB If we use the standard controllability/observability rank tests for this particular hydrogen reformer tenth-order state space model, Chen (2012), MATLAB functions "ctrv" and "obsv" will show that the system is neither controllable nor observable. That is caused by numerical difficulties associated with finding the rank of controllability and observability matrices when the system order is relatively high; in this case it is equal to ten. Fortunately, in the hydrogen gas reformer example, the system matrix A is asymptotically stable (has all eigenvalues in the left half of the complex plane) so that we can use the controllability/observability Gramian tests (Zhou and Doyle 1998; Chen 2012), with the controllability and observability Gramians, respectively, defined by the following integral expressions and the algebraic Lyapunov equations (Gajic and Qureshi 1995):

$$W_c = \int_0^\infty e^{At} BB^T e^{A^T t} dt \quad \Leftrightarrow \quad AW_c + W_c A^T + BB^T = 0 \tag{8.5}$$

$$W_o = \int_0^\infty e^{A^T t} C^T C e^{At} dt \quad \Leftrightarrow \quad A^T W_o + W_o A + C^T C = 0 \tag{8.6}$$

According to Chen (2012), the system is controllable if the controllability Gramian is positive definite (has all eigenvalues with positive real parts) and the system is observable if the observability Gramian is positive definite. To find both controllability Gramians, we can use either MATLAB functions "gram" or "lyap" (that solves the corresponding algebraic Lyapunov equations, given in (8.5) and (8.6)). Using these MATLAB functions, we can verify the hydrogen gas reformer is both controllable and observable, and we can proceed with the design of the reduced-order observer and corresponding optimal controller.

8.2 Full- and Reduced-Order Observer and Optimal Controllers

The goal is to design the linear-quadratic optimal observer-driven controller that will maintain $T_{\text{cpox}} = 972$ K and $y_{\text{H}_2}^{\text{an}} = 0.08$ (8.8%) at steady state despite a piecewise constant disturbance coming from the changes of the fuel cell stack current (due to the fact that the fuel cell current changes as $I_{st} = V_{st}/R_L$, where the load R_L represents the load of all users that turn on and off at random time instants).

In the following, we first show how to design a full-order observer and a reduced-order observer and indicate the advantages of the reduced-order observer over the full-order observer. Then, we present an integral feedback controller design (needed to reject piecewise constant disturbances). The design of the optimal controller driven by the reduced-order observer is done in such a manner that it includes two state variables generated by the integral feedback controller. In addition, a feedforward controller to offset for the impact of the input disturbance is obtained through an optimization procedure. The corresponding simulation results are presented in Sect. 8.3.

8.2.1 Full-Order Observer Design

It is known (see, e.g., Sinha 2007; Chen 2012) that the full-order observer for (8.4) can be designed as follows:

$$\begin{aligned}\frac{\delta\hat{x}(t)}{dt} &= A\delta\hat{x}(t) + B\delta u(t) + Gw(t) + K(\delta y(t) - \delta\hat{y}(t))\\ &= (A - KC)\delta\hat{x}(t) + B\delta u(t) + Gw(t) + K\delta y(t)\\ \delta\hat{y}(t) &= \delta\hat{z}(t) = C\delta\hat{x}(t)\end{aligned} \tag{8.7}$$

The observer gain K has to be chosen such that the observer feedback matrix $A - KC$ is asymptotically stable. That can be achieved by choosing the observer eigenvalues and placing them in the left half complex plane. This can be achieved with a help from the MATLAB function "place" that finds the observer gain K such that the eigenvalues are placed in that desired location using the following K = place(A,C, desired_eigenvalues). Note that the location of the system eigenvalues should be chosen such that the closed-loop observer is considerably faster than the closed-loop system (determined by the eigenvalues of the system closed-loop matrix $A - BF$, where F is a linear proportional full-state feedback gain). The observer is implemented using the SIMULINK state space block as a system with one vector input and one vector output, that is, as

$$\begin{aligned}\frac{\delta\hat{x}(t)}{dt} &= (A - KC)\delta\hat{x}(t) + \begin{bmatrix} B & G & K \end{bmatrix}\begin{bmatrix} \delta u(t) \\ w(t) \\ y(t) \end{bmatrix}\\ \hat{y}(t) &= \delta\hat{z}(t) = C\delta\hat{x}(t)\end{aligned} \tag{8.8}$$

where $(A - KC)$ plays the role of the observer state matrix and $\begin{bmatrix} B & G & K \end{bmatrix}$ is the observer input matrix in the corresponding SIMULINK observer state space block to be implemented in the follow-up of this chapter (see Fig. 8.1).

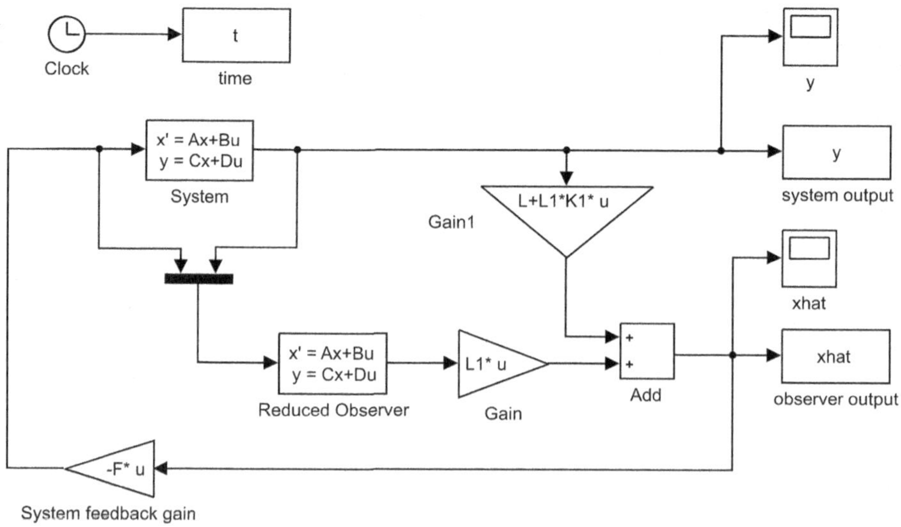

Fig. 8.1 SIMULINK implementation of the reduced-order observer

8.2.2 *Reduced-Order Observer Design*

From (8.4), the formula $\delta y(t) = C\delta x(t)$ gives at all times two algebraic equations for ten state variables, so that we can design a reduced-order observer of dimension eight for the remaining eight state variables Gajic and Lelic (1996). Since the rank of matrix C is $\text{rank}\{C\} = 2$, we can find a matrix C_1 of dimension 8×10 whose rank is equal to 8 such that the augmented matrix

$$\text{rank}\begin{bmatrix} C \\ C_1 \end{bmatrix} = 10 \tag{8.9}$$

has full rank in this case equal to 10.

Introduce a vector $p(t)$ of dimension 8 as

$$\delta p(t) = C_1 \delta x(t) \tag{8.10}$$

and put together the measurement equation and the newly introduced equation (8.10), which produces

$$\begin{bmatrix} \delta y(t) \\ \delta p(t) \end{bmatrix} = \begin{bmatrix} C \\ C_1 \end{bmatrix} \delta x(t) \quad \Rightarrow$$
$$\delta x(t) = \begin{bmatrix} C \\ C_1 \end{bmatrix}^{-1} \begin{bmatrix} \delta y(t) \\ \delta p(t) \end{bmatrix} = [L \quad L_1] \begin{bmatrix} \delta y(t) \\ \delta p(t) \end{bmatrix} = L\delta y(t) + L_1 \delta p(t) \tag{8.11a}$$

$$\begin{bmatrix} C \\ C_1 \end{bmatrix} [L \quad L_1] = I_n = \begin{bmatrix} CL & 0 \\ 0 & C_1L_1 \end{bmatrix} = \begin{bmatrix} I_2 & 0 \\ 0 & I_8 \end{bmatrix} \tag{8.11b}$$

An estimate of $\delta x(t)$ can be obtained from (8.11) as

$$\delta \hat{x}(t) = L\delta y(t) + L_1 \delta \hat{p}(t) \tag{8.12}$$

where $\delta y(t)$ are known measurements and $\delta \hat{p}(t)$ has to be estimated using a reduced-order observer of dimension eight. However, it can be shown that the design of an observer for $\delta \hat{p}(t)$ will require differentiation of the system measurements $\delta y(t)$, which is an undesirably operation. This can be avoided by introducing a change of variables as

$$\delta \hat{q}(t) = \delta \hat{p}(t) - K_1 \delta y(t) \tag{8.13}$$

where K_1 plays the role of the reduced-order observer gain. An observer for $\delta \hat{q}(t)$ can be constructed as a system driven by the system input, disturbance term, and system measurements. It can be shown after some calculations that the reduced-order observer has the form

$$\frac{d\delta \hat{q}(t)}{dt} = A_q \delta \hat{q}(t) + B_q \delta u(t) + G_q w(t) + K_q \delta y(t) \tag{8.14a}$$

where

$$\begin{aligned} A_q &= C_1AL_1 - K_1CAL_1 = (C_1 - K_1C)AL_1 \\ B_q &= C_1B - K_1CB = (C_1 - K_1C)B \\ G_q &= C_1G - K_1CG = (C_1 - K_1C)G \\ K_q &= C_1AL_1K_1 + C_1AL - K_1CAL - K_1CAL_1K_1 = (C_1 - K_1C)A(L + L_1K_1) \end{aligned} \tag{8.14b}$$

The matrix C_1 needed for the design of the reduced-order observer, chosen to make the rank of the augmented matrix given below equal to eight, was taken as

$$C_1 = [\,0_{8\times 2} \quad I_8\,], \qquad \text{rank}\begin{bmatrix} C \\ C_1 \end{bmatrix} = 10 \tag{8.15}$$

The matrix K_1 is chosen to place the closed-loop eigenvalues of the reduced-order observer, the eigenvalues of the matrix A_q, in the desired locations. It can be shown by using the Popov-Belevitch-Hautus eigenvalue test (Chen 2012) that if the original system is observable, that is, the pair (A, C) is observable, then, the pair (C_1AL_1, CAL_1) is also observable, so that the matrix K_1 can be found to place the closed-loop eigenvalues of A_q in the desired locations.

Using the eight-order reduced-order observer (8.14) estimates $\delta\hat{q}(t)$ and formulas (8.11)–(8.12), the optimal estimates of the original state variables $\delta\hat{x}(t)$ are obtained as

$$\delta\hat{x}(t) = L_1\delta\hat{q}(t) + (L + L_1K_1)\delta y(t) \tag{8.16}$$

The estimated output is given by

$$\delta\hat{y}(t) = C\delta\hat{x}(t) \tag{8.17}$$

The simulation results that will clearly show the advantages of the reduced-order observer over the full-order observer will be presented in Sect. 8.3. In addition to producing better simulation results than the full-order observers, the reduced-order observers have implementation advantages since they are simple to build (they are lower order than the full-order observers), and as such they process information faster and more accurately. The reduced-order observers also require the smaller number of feedback loops needed for feedback control that uses the observed (estimated) state variables. The reduced-order observer SIMULINK block diagram is presented in Fig. 8.1.

8.2.3 *Optimal Linear-Quadratic Integral Feedback Controller*

Since the disturbance is a piecewise constant (see Fig. 8.3), we design a feedback loop with an integrator in it (Khalil 2002, Sinha 2007). The integrator is placed such that it is fed with the error term that represents the difference between the desired and actual system output values. Assuming that the system is asymptotically stable, the error input into integrator must be equal to zero (otherwise the integrator will produce huge feedback signals that will destabilize the system). Introducing new variables that are in fact integrals of the error terms, that is

$$\begin{aligned}\frac{d\sigma_1(t)}{dt} &= T_{cpox} - T_{cpox}^{desired} = \delta z_1(t)\\ \frac{d\sigma_2(t)}{dt} &= y_{H_2} - y_{H_2}^{desired} = \delta z_2(t)\end{aligned} \tag{8.18}$$

We form the augmented system

$$\begin{aligned}\begin{bmatrix}\dfrac{d\delta x(t)}{dt}\\ \dfrac{d\sigma}{dt}\end{bmatrix} &= \begin{bmatrix}A & 0\\ C & 0\end{bmatrix}\begin{bmatrix}\delta x(t)\\ \sigma(t)\end{bmatrix} + \begin{bmatrix}B\\ 0\end{bmatrix}\delta u(t) + \begin{bmatrix}G\\ 0\end{bmatrix}\delta w(t)\\ \frac{d\delta x_{aug}(t)}{dt} &= A_{aug}\delta x_{aug}(t) + B_{aug}\delta u(t) + G_{aug}\delta w(t)\end{aligned} \tag{8.19}$$

where

$$\begin{aligned}\sigma(t) = \begin{bmatrix}\sigma_1(t)\\ \sigma_2(t)\end{bmatrix}, \quad \delta x_{aug}(t) = \begin{bmatrix}\delta x(t)\\ \sigma(t)\end{bmatrix}, \quad A_{aug} = \begin{bmatrix}A & 0\\ C & 0\end{bmatrix}\\ B_{aug} = \begin{bmatrix}B\\ 0\end{bmatrix}, \quad G_{aug} = \begin{bmatrix}G\\ 0\end{bmatrix}\end{aligned} \tag{8.20}$$

The optimal control law will be obtained by minimizing a quadratic performance criterion that represents the "squares" of the tracking errors and the "square" of the control input deviation from its nominal steady-state operating points. The corresponding quadratic performance criterion is defined by

$$\begin{aligned}J &= \frac{1}{2}\int_0^\infty \left[\delta x^T(t)C^TQ_zC\delta x(t) + \sigma^T(t)Q_i\sigma(t) + \delta u^T(t)R\delta u(t)\right]dt\\ &= \frac{1}{2}\int_0^\infty \left[\delta x_{aug}^T(t)\begin{bmatrix}C^TQ_zC & 0\\ 0 & Q_i\end{bmatrix}\delta x_{aug}(t) + \delta u^T(t)R\delta u(t)\right]dt\\ &= \frac{1}{2}\int_0^\infty \left[\delta x_{aug}^T(t)Q_{aug}\delta x_{aug}(t) + \delta u^T(t)R\delta u(t)\right]dt\end{aligned} \tag{8.21}$$

The weighted matrices Q_z and Q_i are symmetric and positive semidefinite (most often diagonal), that is, $Q_z = Q_z^T \geq 0$ and $Q_i = Q_i^T \geq 0$, and the matrix R is symmetric and positive definite (the problem solution will require its invertibility), that is, $R = R^T > 0$.

It is shown in Appendix 8.1 that the feedback optimal control strategy, obtained via minimization of the quadratic performance criterion (8.21) along the trajectories of dynamic system (8.19), is given by

$$\delta u(\delta x(t), \delta w(t)) = -F_{aug}\delta x_{aug}(t) + F_w \delta w(t) = -F_{aug}\begin{bmatrix} \delta x(t) \\ \sigma(t) \end{bmatrix} + F_w \delta w(t)$$
$$= -F_x \delta x(t) - F_i \sigma(t) + F_w \delta w(t) = -F_x \delta x(t) - F_i C \int_0^\infty \delta x(t) dt + F_w \delta w(t) \tag{8.22}$$

where

$$F_{aug} = [F_x \quad F_i] = R^{-1} B_{aug}^T P_{aug} \tag{8.23}$$

with P_{aug} representing the positive semidefinite stabilizing solution of the algebraic Riccati equation given by

$$A_{aug}^T P_{aug} + P_{aug} A_{aug} + Q_{aug} - P_{aug} S_{aug} P_{aug} = 0, \quad S_{aug} = B_{aug} R^{-1} B_{aug}^T \tag{8.24}$$

and

$$F_w = R^{-1} B_{aug}^T \left(A_{aug} - S_{aug} P_{aug}\right)^{-T} P_{aug} G_{aug} \tag{8.25}$$

Comment 8.2 It should be emphasized that the terms $G_{aug}\delta w(t)$ in (8.19) and $F_w \delta w(t)$ in (8.22) were not present in the optimization problem considered in Pukrushpan et al. (2004a, b). Hence, our problem formulation and its solution are more rigorous and more complete than the corresponding optimization problem considered in Pukrushpan et al. (2004a, b). Moreover, in Pukrushpan et al. (2004a, b), the full-order observer was used only. Here, we also show that the reduced-order observer produces even better results than the full-order observer for the corresponding observer-driven optimal controller.

Near-optimal linear-quadratic control of the natural gas hydrogen reformer model coupled to the eight-order PEM fuel cell model of Prukrushpan (2004a) via the method of singular perturbations has been recently considered in Nazem-Zadeh and Hamidi-Beheshti (2017).

8.3 Simulation Results

In this section we demonstrate the efficiency of the reduced-order observer-based controller considered in Sect. 8.2 and compare it with the corresponding full-order observer-based controller. We assume that the initial disturbance (current) value is 100 A and that at 10 s the disturbance jumps to 150 A. The simulation results are presented for the time interval of 20 s.

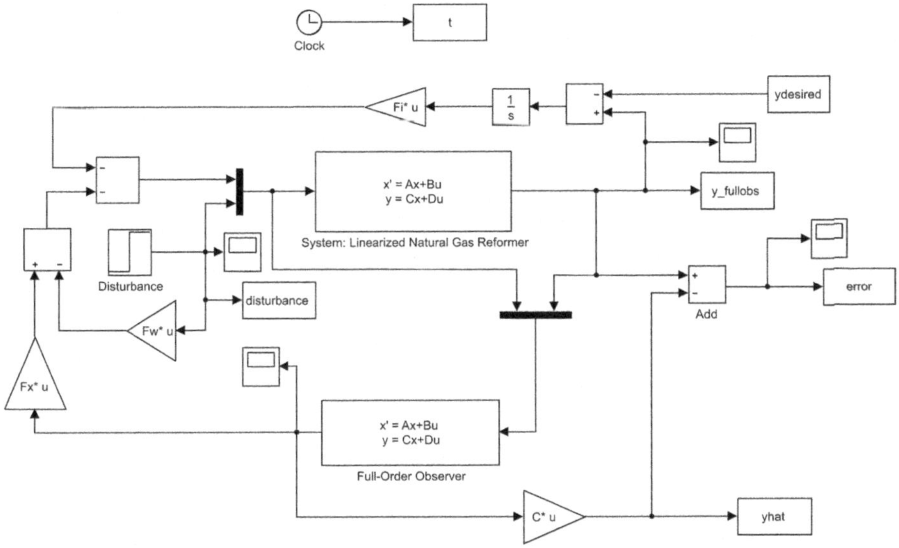

Fig. 8.2 SIMULINK block diagram of the considered controller with the full-order observer

The SIMULINK block diagram of the problem considered in this section is presented in Fig. 8.2 with the full-order observer. The same block diagram can be used with the reduced order observer presented in Fig. 8.1, with the full-order observer in Fig. 8.2 replaced by the reduced-order observer from Fig. 8.1.

The system initial conditions are randomly chosen, and they are kept the same for both the full- and reduced-order observers. They are given by

$$\delta x(0) = [2 \quad 1 \quad 2 \quad 1.5 \quad 2.9 \quad 2.7 \quad 1.5 \quad 2.31 \quad 1.19 \quad 2.5]^T$$

There are no general guidelines how to choose the observer initial conditions. Some authors recommend to set them using the least-square method (Johnson 1988, Stefani et al. 2002) and some to simply set them to zero. In our simulation we will present the results for both cases.

The observer initial condition are chosen via the least-square solution and the Penrose generalized inverse of the measurement equation $\delta y(0) = C\delta x(0)$, which leads to (Johnson 1988, Stefani et al. 2002)

$$\delta\hat{x}(0) = \left(C^T C\right)^{-1} CT\delta y(0) \tag{8.26}$$

The corresponding observer initial conditions obtained via formula (8.26) are

$$\delta\hat{x}(0) = [2.0000 \quad 0.8165 \quad -0.0723 \quad 0 \quad 0 \quad 0 \quad 0 \quad 0 \quad 0 \quad 0]$$

The initial conditions for the reduced-order observer are chosen also via the least-square solution of (8.16) at $t = 0$ (Radisavljevic-Gajic 2015c), that is

$$\delta\hat{x}(0) = L_1\delta\hat{q}(0) + (L + L_1K_1)\delta y(0) \tag{8.27}$$

which leads to

$$\delta\hat{q}(0) = \left(L_1^T L_1\right)^{-1} L_1^T \left[\left(C^T C\right)^{-1} C^T - (L + L_1 K_1)\right] \delta y(0) \tag{8.28}$$

Formula (8.28) produces

$$\hat{q}(0) = [-9.0 \quad -765.5 \quad 0.2 \quad -157.9 \quad 1732.8 \quad 1627.2 \quad -16.6 \quad -23.9]$$

The observer closed-loop eigenvalues are placed at

$$\lambda_{full}^{desired} = [-10 \quad -16 \quad -8 \quad -9 \quad -12 \quad -17 \quad -15 \quad -11 \quad -14 \quad -13]$$

The reduced-order observer closed-loop eigenvalues are placed close to the corresponding full-order observer eigenvalues as

$$\lambda_{reduced}^{desired} = [-13 \quad -12 \quad -15 \quad -16 \quad -11 \quad -10 \quad -14 \quad -9]$$

For the values of the weighted matrices in the performance criterion (8.21), we use the same values as the ones used in Pukrushpan et al. (2004a), which can be found in page 126 of Pukrushpan et al. (2004a), formula (7.14). They are given by

$$Q_z = \begin{bmatrix} 80 & 0 \\ 0 & 1100 \end{bmatrix}, \qquad Q_\sigma = \begin{bmatrix} 150 & 0 \\ 0 & 100 \end{bmatrix}, \qquad R = \begin{bmatrix} 100 & 0 \\ 0 & 120 \end{bmatrix}$$

The corresponding optimal feedback gain can be evaluated by using the MATLAB function "lqr" (linear-quadratic regulator) as follows F = lqr(Aug,Baug,Qaug, R), where the matrix Q_{aug} is the block diagonal matrix: Q = [C′*Qz*C zeros(10,2); zeros(2,10) Qsigma].

Comment 8.3 Note that the observer (full-order or reduced-order) is supposed to be much faster than the closed-loop system, namely, the closed-loop observer eigenvalues are supposed to be placed in the complex plane left from the closed-loop system eigenvalues. The closed-loop system (hydrogen gas reformer) eigenvalues (based on the choice of the weighted matrices) were clustered in two groups, with seven of them being close to the imaginary axis (slow eigenvalues) and three of them being fast and located far to the left from imaginary axis, including one of them located at -661. Hence, to satisfy the requirement that the observer is much faster than the system, one would have to place the observer eigenvalues at -3000 or so and even further to the left. Such an observer with extremely fast dynamics will

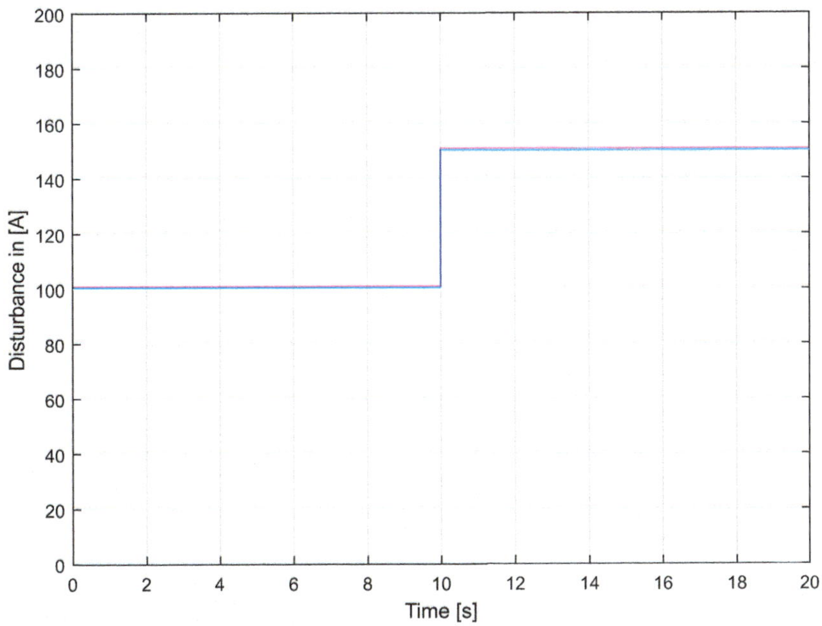

Fig. 8.3 Variations of the fuel cell disturbance (current)

require extremely broad bandwidth, huge observer gains, and be prone to noise. However, it can be seen from the simulation results presented that the closed-loop eigenvalues chosen for both the full- and reduced-order observers produce pretty good results for this particular model of the hydrogen gas reformer. If one intends to have this theoretical result satisfied in practice (the observer should be much faster than the system), one should use different weighted matrices Q_z, Q_i, and R that will provide appropriate placement of the gas reformer closed-loop eigenvalues or even to study the hydrogen gas reformer in two time scales, slow and fast, and design corresponding slow and fast time scales controllers and observers. These will be interesting topics for future research.

The obtained results using the proposed controller defined in (8.22), (8.23), (8.24), (8.25), and (8.26) with the full-order observer with the observer initial conditions obtained using the least-square method are presented in Fig. 8.4, and the waveform of the disturbance current is presented in Fig. 8.3.

The corresponding simulation results when both the full- and reduced-order observer initial conditions are set to zero show the similar responses to those presented in Figs. 8.4 and 8.5 so that in this case also, the reduced-order observer has a better transient response for the observed system output than the full-order observer.

The simulation results obtained for the temperature variations from its nominal value (upper curve) and the hydrogen molar fraction variations (lower curve) are presented in Fig. 8.4. It can be seen from these figures that despite of large variations

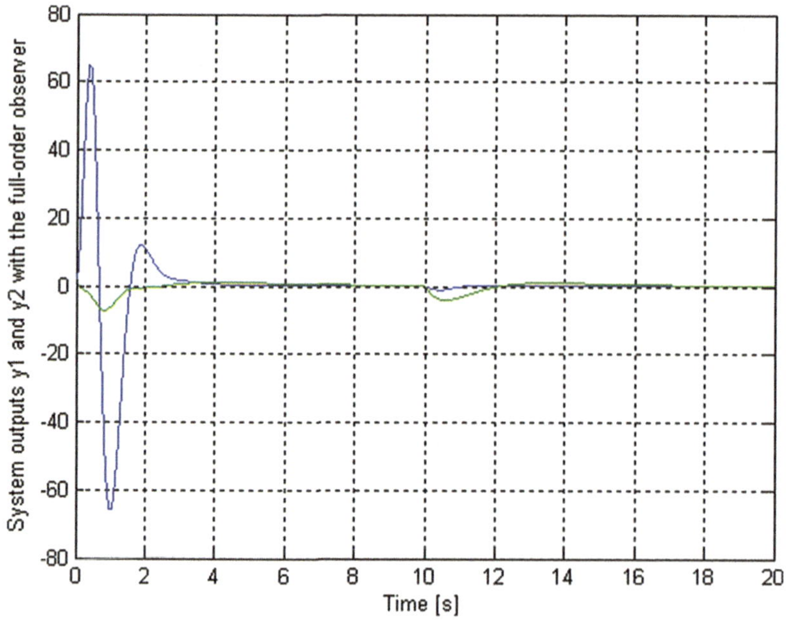

Fig. 8.4 The temperature variations (blue curve) and the hydrogen molar fraction variations (green curve) around their nominal values using the full-order observer. The goal is that the outputs take the zero values. The observer initial conditions are obtained from (8.26)

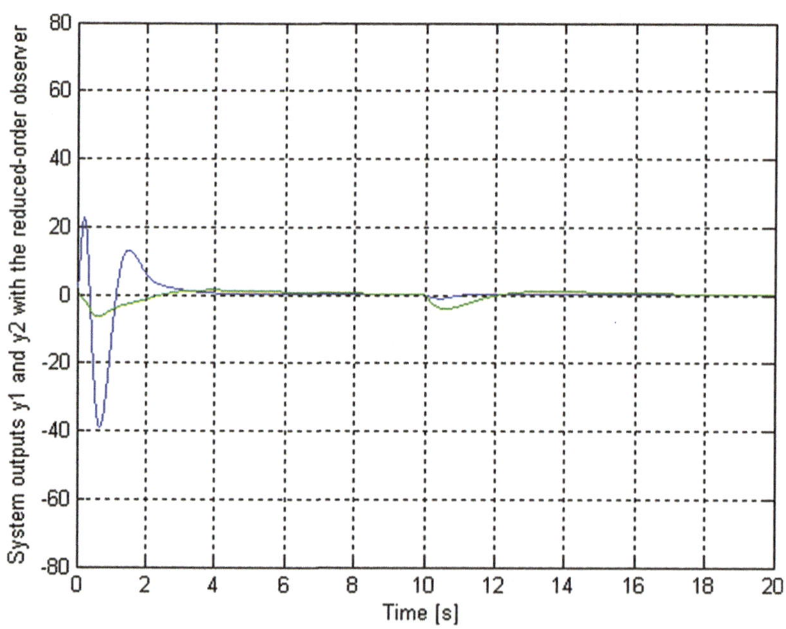

Fig. 8.5 The temperature variations (blue curve) and the hydrogen molar fraction variations (green curve) around their nominal values using the reduced-order observer. The goal is that the outputs take zero values. The observer initial conditions are obtained from (8.28)

in the stack current, the output (controlled) variables are only slightly changed at the disturbance jump time, and the corresponding ripples disappear quickly, within a few seconds. Note that variations about nominal values can be negative even for physical quantities that are positive at all times. In Fig. 8.5 we present the corresponding results for the reduced-order observer with its initial values obtained via the least-square method. It can be noticed that the reduced-order observer produces better results as far as the initial transient is concerned (which is the effect of the observer efficiency) and that both observers have similar performances as far as the rejection of the constant disturbance is concerned (which is contributed by the controller efficiency).

In summary, we have designed an optimal reduced-order observer-based controller that has both proportional and integral feedback loops from the state variables and the proportional feed-forward controller from the fuel cell current (disturbance). It has been observed that it is beneficial to use the reduced-order observer instead of the full-order observer for this particular application. The results are promising. They were derived analytically and verified via simulation for the linearized model of the considered hydrogen gas reformer. Studying the hydrogen gas reformer in two time scales, slow and fast, and designing corresponding slow and fast time scales controllers and observers will be also interesting topics for future research.

8.4 Notes

Presentation of material in chapter is based on the conference paper Radisavljevic-Gajic and Rose (2015). Permission for the use of such material in this research monograph was granted to us by the American Society of Mechanical Engineers (ASME) for the conference paper Radisavljevic-Gajic and Rose (2015) presented at the 2015 ASME Dynamic Systems and Control Conference.

Appendix 8.1

To minimize with respect to the control variable $\delta u(t)$ the quadratic performance criterion

$$J=\frac{1}{2}\int_{0}^{\infty}\left[\delta x_{aug}^{T}(t)Q_{aug}\delta x_{aug}(t)+\delta u^{T}(t)R\delta u(t)\right]dt \tag{8.A.1}$$

along trajectories of the dynamic system

$$\frac{d\delta x_{aug}(t)}{dt} = A_{aug}\delta x_{aug}(t) + B_{aug}\delta u(t) + G_{aug}\delta w(t) \tag{8.A.2}$$

we first form the Hamiltonian, Kirk (2004)

$$H = \frac{1}{2}\left(\delta x_{aug}^T Q_{aug}\delta x_{aug} + \delta u^T R\delta u\right) + \delta p^T\left(A_{aug}\delta x_{aug} + B_{aug}\delta u + G_{aug}\delta w\right) \tag{8.A.3}$$

and then take the partial derivatives which provide the necessary condition for the minimum, Kirk (2004)

$$\frac{d\delta x_{aug}}{dt} = \frac{\partial H}{\partial \delta p^T} = A_{aug}\delta x_{aug} + B_{aug}\delta u + G_{aug}\delta w \tag{8.A.4}$$

$$\frac{d\delta p}{dt} = -\frac{\partial H}{\partial \delta x} = -Q_{aug}\delta x_{aug} - A_{aug}\delta p \tag{8.A.5}$$

$$0 = \frac{\partial H}{\partial \delta u} = R\delta u + B_{aug}^T\delta p \quad \Rightarrow \quad \delta u = -R^{-1}B_{aug}^T\delta p \tag{8.A.6}$$

Substituting δu from (8.A.6) into (8.A.4), we get the system of differential equations in the form

$$\frac{d\delta x_{aug}}{dt} = A_{aug}\delta x_{aug} - S_{aug}\delta p + G_{aug}\delta w, \quad S_{aug} = B_{aug}R^{-1}B_{aug}^T \tag{8.A.7}$$

$$\frac{d\delta p}{dt} = -\frac{\partial H}{\partial \delta x} = -Q_{aug}\delta x_{aug} - A_{aug}^T\delta p \tag{8.A.8}$$

The system (8.A.7) and (8.A.8) can be solved by looking for its solution in the form

$$\delta p(t) = P_{aug}x_{aug}(t) + \gamma \tag{8.A.9}$$

where K and γ are, respectively, a constant matrix and a constant vector to be determined. Taking the derivative of (8.A.9) and using (8.A.7) and (8.A.8), we obtain

$$\begin{aligned} &-\left(A_{aug}^T P_{aug} + P_{aug}A_{aug} + Q_{aug} - P_{aug}S_{aug}P_{aug}\right)\delta x(t) \\ &= \left(A_{aug}^T\gamma - P_{aug}S_{aug}\gamma + P_{aug}G_{aug}\delta w(t)\right) \end{aligned} \tag{8.A.10}$$

From this equality we have

$$A_{aug}^T P_{aug} + P_{aug}A_{aug} + Q_{aug} - P_{aug}S_{aug}P_{aug} = 0 \tag{8.A.11}$$

and

$$\begin{aligned} &A_{aug}^T\gamma - P_{aug}S_{aug}\gamma + P_{aug}G_{aug}\delta w(t) = 0 \\ &\Rightarrow \gamma = \left(A_{aug} - S_{aug}P_{aug}\right)^{-T} P_{aug}G_{aug}\delta w(t) \end{aligned} \tag{8.A.12}$$

Equation (8.A.11) represents the well-known algebraic Riccati equation. Its unique positive semidefinite solution that stabilizes the closed-loop system exists under the conditions that the pairs (A_{aug},B_{aug}) and (A_{aug},Chol(Q_{aug})), where "Chol" stands for the Cholesky decomposition (Golub and Van Loan 2012), are, respectively, stabilizable (controllable) and detectible (observable). It is interesting to observe that stabilizability (controllability) of the pair (A_{aug},B_{aug}) can be expressed in terms of stabilizability (controllability) of the original pair (A, B) (Smith and Davison 1972). That result is stated in Lemma 8.1. Similarly, detectability (observability) of the augmented system can be expressed in terms of detectability (observability) of the original system (Smith and Davison 1972; see Lemma 8.2).

Lemma 8.1 The pair (A_{aug},B_{aug}) is stabilizable (controllable) if the pair (A, B) is stabilizable (controllable) and the matrix $\begin{bmatrix} A & B \\ C & 0 \end{bmatrix}$ has full rank.

Lemma 8.2 The pair $\left(A_{aug}, \sqrt{Q_{aug}}\right)$ is detectible (observable) if the pair (A, C) is detectible (observable) and the matrix $\begin{bmatrix} A & C^TQ_xC \\ C & 0 \end{bmatrix}$ has full rank.

Having obtained P_{aug} from (8.A.11) and γ from (8.A.12), the optimal control $\delta u(t)$ that minimizes (8.A.1) along trajectories of (8.A.2) can be found from (8.A.6) and (8. A.9) as

$$\begin{aligned} \delta u(\delta x(t), \sigma(t), \delta w(t)) &= -R^{-1}B_{aug}^T\delta p = -R^{-1}B_{aug}^T\left(P_{aug}x_{aug}(t) + \gamma\right) \\ &= -F_{aug}\delta x_{aug}(t) + F_w\delta w(t) = -F_{aug}\begin{bmatrix} \delta x(t) \\ \sigma(t) \end{bmatrix} + F_w\delta w(t) \\ &= -F_x\delta x(t) - F_i\sigma(t) + F_w\delta w(t) \\ &= -F_x\delta x(t) - F_iC\int_0^\infty \delta x(t)dt + F_w\delta w(t) \end{aligned} \tag{8.A.13}$$

where

$$F_{aug} = [F_x \quad F_i] = R^{-1}B_{aug}^TP_{aug} \tag{8.A.14}$$

and

$$F_w = R^{-1}B_{aug}^T\left(A_{aug} - S_{aug}P_{aug}\right)^{-T}P_{aug}G_{aug} \tag{8.A.15}$$

Chapter 9
Extensions to Multi-stages and Multi-time Scales

This monograph has presented the recent research results of the authors on design of multistage and multi-time-scale feedback controllers for linear, continuous- and discrete-time, time-invariant, linear dynamic systems. This research is particularly important for large-scale dynamic systems composed of several subsystems and/or large linear systems operating in several time scales. In addition, it is important for higher-order linear dynamic systems whose design of feedback controllers can be facilitated by artificially introducing subsystems of much lower order than the original system.

Applying the proposed methodology, different types of linear feedback controllers can be designed for different parts of the system, or in general, only desired partial control of some parts of the system can be achieved. Such controllers designed for particular parts of the system ought to be the most appropriate controllers for those subsystems. This important feature of multistage design of linear feedback controllers is not present in any other linear controller design technique.

The presented results are only the first steps in this important field. Many theoretical problems about the design of multistage feedback controllers are open for future research, especially for fourth- and higher-stage designs, and in general, for large linear systems composed of N subsystems. Some of the potential research problems have been already outlined in the book, including the design of multistage and multi-time scale Kalman filter in both continuous- and discrete-time domains, which we consider very important due to very broad applications of the Kalman filter in almost all fields of sciences and engineering. The use of finite-time linear-quadratic optimal controllers for local subsystems is possible, even though it has not been considered in this book. Extensions of the presented methodology to linear time-varying control systems might be very challenging, if possible at all.

We have demonstrated applications of two- and three-stage linear feedback controller designs on the examples of proton exchange membrane fuel cells and related energy systems, including linear dynamic systems that operate in two- and three-time scales. There are numerous potential applications of the presented

V. Radisavljević-Gajić et al., *Multi-Stage and Multi-Time Scale Feedback Control of Linear Systems with Applications to Fuel Cells*, Mechanical Engineering Series,
https://doi.org/10.1007/978-3-030-10389-7_9

feedback design methodology to other real physical linear or linearized mathematical models of dynamic systems.

At the end of this section, we will present systematic summaries of the derived formulas for linear feedback controller designs for three and four stages, as a motivation and guidelines for derivations of the general N-stage design of linear feedback controllers.

9.1 Extensions to Multi-stage Multi-time Scale Linear Systems

Linear time-invariant systems to be subjected to the multi-stage feedback design should be first appropriately partitioned and their subsystems identified. The partitioning can be done using several criteria: (a) based on the physical nature of subsystems parts (system natural decomposition), (b) according to the conditions that must be satisfied such that the partitioned system is feasible for the multistage feedback design, (c) based on mathematical conditions that must be satisfied to solve the corresponding design equations, and (d) control needs (which parts of the system should be independently controlled via local feedback controllers) or based on grouping of the state space variables such that the subsystems satisfy control-oriented assumptions needed for the design of local controllers such as controllability (stabilizability), observability (detectability) and similar requirements).

The partitioned continuous-time, time-invariant, large-scale (complex) linear system is defined by

$$\begin{bmatrix} \frac{dx_1(t)}{dt} \\ \frac{dx_2(t)}{dt} \\ \vdots \\ \vdots \\ \frac{dx_N(t)}{dt} \end{bmatrix} = \begin{bmatrix} A_{11} & A_{12} & \cdots & \cdots & A_{1N} \\ A_{21} & A_{22} & A_{23} & \cdots & A_{2N} \\ \vdots & \ddots & \ddots & \ddots & \vdots \\ \vdots & \vdots & \ddots & \ddots & A_{N-1,N} \\ A_{N1} & A_{N2} & \cdots & A_{N,N-1} & A_{NN} \end{bmatrix} \begin{bmatrix} x_1(t) \\ x_2(t) \\ \vdots \\ \vdots \\ x_N(t) \end{bmatrix} + \begin{bmatrix} B_{11} \\ B_{22} \\ \vdots \\ \vdots \\ B_{NN} \end{bmatrix} u(t) \tag{9.1}$$

where $x_i(t) \in R^{n_i}$, $x(t) \in R^n$, $n = n_1 + n_2 + \cdots + n_N$, and are state variables, $u(t) \in R^m$ is the control input vector, and A_{ij} and B_{ii}, $i,j = 1,2,\ldots,N$, are constant matrices of appropriate dimensions. Matrices A_{ii} define subsystems of dimensions n_i corresponding to state variables $x_i(t)$. Matrices A_{ij}, $i \neq j$ define couplings between the subsystems. N represents the number of subsystems in the system.

The special structure of (9.1) was studied in Gajic et al. (2009) in the context of weakly coupled linear control systems, where the following change of state variables was used

$$\eta_i(t) = x_i(t) + \sum_{j=1, j \neq i}^{N} L_{ij} x_j(t), \quad i = 1,2, \ldots, N \tag{9.2}$$

with L_{ij} satisfying a set of algebraic equations

$$\begin{aligned} &L_{ij}A_{jj} - A_{ii}L_{ij} + A_{ij} + \sum_{k=1,\, k \neq i,j}^{N} L_{ik}A_{kj} - \left(\sum_{k=1,\, k \neq i,j}^{N} L_{ik}A_{kj} \right) L_{ij} = 0 \\ &i,j = 1,2, \ldots, N, \quad i \neq j \end{aligned} \tag{9.3}$$

Transformation (9.2) and (9.3) applied to (9.1) leads to a pure *block-diagonal form* for the system matrix in the new coordinates. In the work of Gajic et al. (2009), the matrices A_{kj} were $A_{kj} = O(\varepsilon), \ k \neq j$ so that nonlinear algebraic equation (9.3) were solved iteratively using the fixed-point iterations as sets of linear Sylvester-type algebraic equations, whose solutions existed in each iteration under the assumption that matrices A_{jj} and $A_{ii}, i,j = 1,2, \ldots, N$, have no eigenvalues in common. Up to our best knowledge, the general structure (when matrices A_{ii} have dimensions $n_i \times n_i$ and $A_{kj} = O(1), \ k \neq j$) of the algebraic equation (9.3) has not been studied in any research paper so that the existence of solutions and numerical methods for finding those solutions remain unknown. Based on what we have presented and learnt from Chaps. 2, 3, 4, 5 and 6, the pure block-diagonal structure obtained in Gajic et al. (2009) by applying (9.2) and (9.3) to (9.1) is more than what we need for the development of the multistage linear feedback controller design. Note that algebraic equations in (9.3) are nonlinear quadratic-type algebraic equations.

In the process of extending the two-, three-, and four-stage feedback controller designs to multistages, using the methodology presented in Chaps. 2, 4, and 6 of this monograph, the first step will be to find a transformation that puts (9.1) into the *upper block triangular form* defined by

$$\begin{bmatrix} \dfrac{dz_1(t)}{dt} \\ \dfrac{dz_2(t)}{dt} \\ \vdots \\ \vdots \\ \dfrac{dz_N(t)}{dt} \end{bmatrix} = \begin{bmatrix} A_{11} & A_{12} & \cdots & \cdots & A_{1N} \\ 0 & A_{22} & A_{23} & \cdots & A_{2N} \\ \vdots & 0 & \ddots & \ddots & \vdots \\ \vdots & \vdots & \ddots & \ddots & A_{N-1,N} \\ 0 & 0 & \cdots & 0 & A_{NN} \end{bmatrix} \begin{bmatrix} z_1(t) \\ z_2(t) \\ \vdots \\ \vdots \\ z_N(t) \end{bmatrix} + \begin{bmatrix} B_1 \\ B_2 \\ \vdots \\ \vdots \\ B_N \end{bmatrix} u(t) \tag{9.4}$$

We expect that the change of state space variables needed to get the form defined in (9.4) will be similar to those given in (9.2) and that the algebraic equations whose solutions facilitate the upper block triangular structure will have the form similar to (9.3).

It is possible that the multi-stage feedback controller design technique can be done more efficiently by converting the original system (9.1) into the following form:

$$\begin{bmatrix} \frac{d\eta_1(t)}{dt} \\ \frac{d\eta_2(t)}{dt} \\ \vdots \\ \frac{d\eta_{N-1}(t)}{dt} \\ \frac{d\eta_N(t)}{dt} \end{bmatrix} = \begin{bmatrix} A_{11} & A_{12} & \cdots & \cdots & A_{1N} \\ A_{21} & A_{22} & A_{23} & \cdots & A_{2N} \\ \vdots & \ddots & \ddots & \ddots & \vdots \\ A_{N-1,1} & A_{N-1,2} & \ddots & \ddots & A_{N-1,N} \\ 0 & 0 & \cdots & 0 & A_{NN} \end{bmatrix} \begin{bmatrix} \eta_1(t) \\ \eta_2(t) \\ \vdots \\ \vdots \\ \eta_N(t) \end{bmatrix} + \begin{bmatrix} B_{11} \\ B_{22} \\ \vdots \\ \vdots \\ B_{NN} \end{bmatrix} v(t) \tag{9.5}$$

This form allows an independent design of a controller for the *N*-th subsystem, for example

$$u(t) = -\Gamma_N \eta_N(t) + v(t) \tag{9.6}$$

In the next step, a state transformation has to be found to facilitate independent controller design for the *(N-1)*-th subsystem. To that end, in the next design stage, a state transformation has to be found that puts the system into the following form

$$\begin{bmatrix} \frac{d\eta_1(t)}{dt} \\ \frac{d\eta_2(t)}{dt} \\ \vdots \\ \frac{d\eta_{N-1}(t)}{dt} \\ \frac{d\eta_N(t)}{dt} \end{bmatrix} = \begin{bmatrix} A_{11} & A_{12} & \cdots & \cdots & A_{1N} \\ A_{21} & A_{22} & A_{23} & \cdots & A_{2N} \\ \vdots & \ddots & \ddots & \ddots & \vdots \\ 0 & 0 & 0 & A_{N-1,N-1} & A_{N-1,N} \\ 0 & 0 & \cdots & 0 & A_{NN} - B_{NN}\Gamma_N \end{bmatrix} \begin{bmatrix} \eta_1(t) \\ \eta_2(t) \\ \vdots \\ \eta_{N-1}(t) \\ \eta_N(t) \end{bmatrix} + \begin{bmatrix} B_{11} \\ B_{22} \\ \vdots \\ B_{N-1,N-1} \\ B_{NN} \end{bmatrix} v(t) \tag{9.7}$$

The procedure has to be continued backward until independent controllers are designed for each subsystems. It is expected that in this process, a sequence of state transformations similar to (9.2) has to be found and a set of coupled nonlinear algebraic equations similar to (9.3) be derived.

The potential future research problem can be broken down into the following research tasks:

Task 9.1: Find the state transformation that puts the model of the original system (9.1) into the form defined in (9.4).

Task 9.2: Study the corresponding set of nonlinear algebraic equations needed for the transformation established in Task 9.1. It is expected that we will get N nonlinear coupled quadratic algebraic equations and $N^2 - N$ linear coupled Sylvester-type algebraic equations. This task is twofold: (a) establish the existence of solutions of the derived algebraic equations, and (b) develop efficient numerical algorithms for solving the corresponding algebraic equations. Note that any solution of these nonlinear algebraic equations can be used for the purpose of the multi-stage design. Hence, establishing conditions for the existence of the unique solution is not necessary for the considered multistage linear feedback controller design.

Task 9.3: Using the subsystem feedback gains, to be obtained by applying the appropriate control system design techniques to the corresponding subsystems, we need to derive the formula for the equivalent feedback gain for the full-state feedback control to be applied to the original system in the original coordinates.

Comment 9.1: In an idealistic case, the subsystems can be defined down to the most elementary subsystems consisting of one-dimensional subsystems (for real eigenvalues) and two-dimensional subsystems (for complex conjugate eigenvalues). This can be achieved by using the well-known QR algorithm (the most efficient algorithm for finding the matrix eigenvalues) (Stewart 1973; Golub and Van Loan 2012) that brings any square matrix in the upper block-diagonal form with the diagonal block having dimensions of one or two. This form is known as the real Schur form (Stewart 1973; Golub and Van Loan 2012). It is given by

$$
\begin{bmatrix} \frac{dz_1(t)}{dt} \\ \frac{dz_2(t)}{dt} \\ \vdots \\ \vdots \\ \frac{dz_S(t)}{dt} \end{bmatrix} = \begin{bmatrix} \lambda_1 & A_{12} & \cdots & \cdots & A_{1S} \\ 0 & \lambda_2 & A_{23} & \cdots & A_{2S} \\ \vdots & 0 & \ddots & \ddots & \vdots \\ \vdots & \vdots & \ddots & \begin{bmatrix} \alpha_{S-1} & \beta_{S-1} \\ -\beta_{S-1} & \alpha_{S-1} \end{bmatrix} & A_{S-1,S} \\ 0 & 0 & \cdots & 0 & \begin{bmatrix} \alpha_S & \beta_S \\ -\beta_S & \alpha_S \end{bmatrix} \end{bmatrix} \begin{bmatrix} z_1(t) \\ z_2(t) \\ \vdots \\ \vdots \\ z_S(t) \end{bmatrix} + \begin{bmatrix} B_1 \\ B_2 \\ \vdots \\ \vdots \\ B_S \end{bmatrix} u(t) \tag{9.8}
$$

where S stands for the number of distinct eigenvalues. We have assumed that the complex eigenvalues are given by $\lambda_i = \alpha_i \pm j\beta_i$. This structure defined in (9.8) should be studied from the multistage feedback controller design point of view. However, since the form (9.8) is obtained numerically, it is unrealistic to expect that all conditions needed for the multistage feedback design will be satisfied. Due to the fascinating fact that we might be able to design independently linear feedback controllers for almost every single eigenvalue, including pairs of complex conjugate eigenvalues, it is a sufficient reason to consider this research problem as well in the near future.

At the end of this section, we present two appendices, Appendixes 9.1 and 9.2 that summarize in a systematic manner the design phases and steps used in the development of three- and four-stage linear feedback controllers. We hope that they will provide some useful ideas in the development of the general multistage linear feedback controllers. In the case of the three-stage linear feedback controller design, in Appendix 9.1, we provide also the corresponding summary for three-time scale (singularly perturbed) linear dynamic systems.

9.2 Multi-stage Feedback Design for Multi-time Scale Systems

As indicated throughout this research monograph, several multi-time scales are present in many real physical systems that have components of different nature (electrical, mechanical, chemical, thermodynamic, etc.) (Kokotovic et al. 1999; Naidu and Calise 2001; Gajic and Lim 2001; Kuehn 2015).

The general structure of a multi-time scale continuous-time, time-invariant linear system is defined by

$$\begin{bmatrix} \frac{dx_1(t)}{dt} \\ \varepsilon_1 \frac{dx_2(t)}{dt} \\ \varepsilon_2 \frac{dx_2(t)}{dt} \\ \vdots \\ \varepsilon_{n-1} \frac{dx_N(t)}{dt} \end{bmatrix} = \begin{bmatrix} A_{11} & A_{12} & \cdots & \cdots & A_{1N} \\ A_{21} & A_{22} & A_{23} & \cdots & A_{2N} \\ \vdots & \ddots & \ddots & \ddots & \vdots \\ \vdots & \vdots & \ddots & \ddots & A_{N-1,N} \\ A_{N1} & A_{N2} & \cdots & A_{N,N-1} & A_{NN} \end{bmatrix} \begin{bmatrix} x_1(t) \\ x_2(t) \\ \vdots \\ \vdots \\ x_N(t) \end{bmatrix} + \begin{bmatrix} B_{11} \\ B_{22} \\ \vdots \\ \vdots \\ B_{NN} \end{bmatrix} u(t) \tag{9.9}$$

where $1 >> \varepsilon_1 >> \varepsilon_2 >> \cdots >> \varepsilon_{N-1} > 0$ are small positive parameters that indicate the presence of multiple time scales.

The multi-stage multi-time scale feedback controller design will remove numerical ill-conditioning present in multi-time scale systems (Kokotovic et al. 1999; Naidu and Calise 2001) and facilitate design of independent feedback controllers, observers, and filters for this class of linear dynamical systems. Moreover, it is expected that the nonlinear algebraic equations needed for the state transformation and the recovery of the equivalent full-state feedback gain in the original coordinates will get simpler forms than for the general linear time-invariant systems partitioned into subsystems and defined in (9.1). As a matter of fact, it is expected that the observations made in Sects. 2.4 and 4.3 will be valid for this research task so that all needed algebraic equations due to the presence of small parameters will be solved iteratively as systems linear matrix equations and systems of Sylvester algebraic equations.

The methodology developed in the continuous-time domain for multistage linear feedback design of control systems could be, and should be, extended to the corresponding discrete-time domain counterparts, both in the general case and in the case of multi-time scale linear control systems.

9.3 Multi-stage Feedback Design for Other Classes of Systems

It is possible to study some classes of distributed parameter systems, flexible space structures, and nonclassically damped second-order linear systems since their dynamic models can be represented using subsystems that are mutually coupled like in our basic structure defined in (9.1). Linearized models of distributed parameter systems (described by partial differential equations) can be represented in the modal coordinates by infinite sets of second-order ordinary differential equations, Meirovich and Baruh (1983), Baruh and Choe (1990). Flexible structures, especially large space flexible structures, can be modeled by infinite series of pure oscillators (having eigenvalues on the imaginary axis) and lightly damped oscillators (having eigenvalues in the stable half complex plane very close to the imaginary axis), Gawronski and Juang (1990), Gawronski (1994, 1998). In both cases, the infinite series can be well approximated by finite series. Another class of systems that can be studied using multistage feedback design is second-order nonclassically damped linear mechanical systems (Radisavljevic-Gajic 2013). These systems are represented by N second-order scalar dynamics systems mutually weakly coupled so that the coupling matrices are $A_{kj} = O(\varepsilon),\ k \neq j$, which will facilitate simplification of the nonlinear algebraic equations needed for the multistage feedback design.

Appendix 9.1: Summary of the Three-Stage Feedback Controller Design (Tables 9.1, 9.2, 9.3 and 9.4)

Table 9.1 Introduction to three-stage continuous-time feedback controller design

Original partitioned system	$\begin{bmatrix}\dot{x}_I\\ \dot{x}_{II}\\ \dot{x}_{III}\end{bmatrix}=\begin{bmatrix}A_{11} & A_{12} & A_{13}\\ A_{21} & A_{22} & A_{23}\\ A_{31} & A_{32} & A_{33}\end{bmatrix}\begin{bmatrix}x_I\\ x_{II}\\ x_{III}\end{bmatrix}+\begin{bmatrix}B_{11}\\ B_{22}\\ B_{33}\end{bmatrix}u, y=\begin{bmatrix}C_{11} & C_{22} & C_{33}\end{bmatrix}\begin{bmatrix}x_I\\ x_{II}\\ x_{III}\end{bmatrix}$	
$\underbrace{\eta_3=L_1x_I+L_2x_{II}+x_{III}}_{transform}$	$\rightarrow$	$\begin{bmatrix}\dot{x}_I\\ \dot{x}_{II}\\ \dot{\eta}_3\end{bmatrix}=\begin{bmatrix}A_{11}-A_{13}L_1 & A_{12}-A_{13}L_2 & A_{13}\\ A_{21}-A_{23}L_1 & A_{22}-A_{23}L_2 & A_{23}\\ f_{31}(L_1,L_2) & f_{32}(L_1,L_2) & A_3\end{bmatrix}\begin{bmatrix}x_I\\ x_{II}\\ \eta_3\end{bmatrix}+\begin{bmatrix}B_{11}\\ B_{22}\\ B_3\end{bmatrix}u$ $A_3=L_1A_{13}+L_2A_{23}+A_{33},\ B_3=L_1B_{11}+L_2B_{22}+B_{33}$
Solve for L_1,L_2		$f_{31}(L_1,L_2)=L_1A_{11}+L_2A_{21}+A_{31}-A_3L_1=0$ $f_{32}(L_1,L_2)=L_1A_{12}+L_2A_{22}+A_{32}-A_3L_2=0$
$\underbrace{\eta_2=L_3x_I+x_{II}}_{transform}$	$\rightarrow$	$\begin{bmatrix}\dot{x}_I\\ \dot{\eta}_2\\ \dot{\eta}_3\end{bmatrix}=\begin{bmatrix}A_1 & A_{12}-A_{13}L_2 & A_{13}\\ f_{21}(L_3) & A_2 & A_{23}+L_3A_{13}\\ 0 & 0 & A_3\end{bmatrix}\begin{bmatrix}x_I\\ \eta_2\\ \eta_3\end{bmatrix}+\begin{bmatrix}B_{11}\\ B_2\\ B_3\end{bmatrix}u$ $A_1=(A_{11}-A_{13}L_1)-(A_{12}-A_{13}L_2)L_3$ $A_2=A_{22}-A_{23}L_2+L_3(A_{12}-A_{13}L_2),\ B_2=B_{22}+L_3B_{11}$
Solve for L_3	$f_{21}(L_3)=L_3(A_{11}-A_{13}L_1)-(A_{22}-A_{23}L_2)L_3-L_3(A_{12}-A_{13}L_2)L_3+(A_{21}-A_{23}L_1)=0$	
Upper triangular form		$\begin{bmatrix}\dot{x}_I\\ \dot{\eta}_2\\ \dot{\eta}_3\end{bmatrix}=\begin{bmatrix}A_1 & A_{12}-A_{13}L_2 & A_{13}\\ 0 & A_2 & A_{23}+L_3A_{13}\\ 0 & 0 & A_3\end{bmatrix}\begin{bmatrix}x_I\\ \eta_2\\ \eta_3\end{bmatrix}+\begin{bmatrix}B_{11}\\ B_2\\ B_3\end{bmatrix}u=\bar{A}\begin{bmatrix}x_I\\ \eta_2\\ \eta_3\end{bmatrix}+\bar{B}u$ $y=(C_{11}-C_{22}L_3-C_{33}L_1)x_I+(C_{22}-C_{33}L_2)\eta_2+C_{33}\eta_3=$ $=C_1x_I+C_2\eta_2+C_{33}\eta_3=\bar{C}\begin{bmatrix}x_I & \eta_2 & \eta_3\end{bmatrix}^T$
Similarity transformation that maps original system in the upper triangular form		$\begin{bmatrix}x_I\\ \eta_2\\ \eta_3\end{bmatrix}=T_1\begin{bmatrix}x_I\\ x_{II}\\ x_{III}\end{bmatrix}=\begin{bmatrix}I & 0 & 0\\ L_3 & I & 0\\ L_1 & L_2 & I\end{bmatrix}\begin{bmatrix}x_I\\ x_{II}\\ x_{III}\end{bmatrix}\begin{bmatrix}x_I\\ x_{II}\\ x_{III}\end{bmatrix}$ $=T_1^{-1}\begin{bmatrix}x_I\\ \eta_2\\ \eta_3\end{bmatrix}=\begin{bmatrix}I & 0 & 0\\ -L_3 & I & 0\\ -L_1+L_2L_3 & -L_2 & I\end{bmatrix}\begin{bmatrix}x_I\\ \eta_2\\ \eta_3\end{bmatrix}$

Table 9.2 Design of three-stage feedback controllers

Upper triangular form	$\begin{bmatrix} \dot{x}_I \\ \dot{\eta}_2 \\ \dot{\eta}_3 \end{bmatrix} = \begin{bmatrix} A_1 & A_{12}-A_{13}L_2 & A_{13} \\ 0 & A_2 & A_{23}+L_3A_{13} \\ 0 & 0 & A_3 \end{bmatrix} \begin{bmatrix} x_I \\ \eta_2 \\ \eta_3 \end{bmatrix} + \begin{bmatrix} B_{11} \\ B_2 \\ B_3 \end{bmatrix} u = \bar{A} \begin{bmatrix} x_I \\ \eta_2 \\ \eta_3 \end{bmatrix} + \bar{B}u$ $A_1 = (A_{11} - A_{13}L_1) - (A_{12} - A_{13}L_2)L_3, A_2 = A_{22} - A_{23}L_2 + L_3(A_{12} - A_{13}L_2), B_2 = B_{22} + L_3B_{11}$ $A_3 = L_1A_{13} + L_2A_{23} + A_{33}B_3 = L_1B_{11} + L_2B_{22} + B_{33}, B_3 = L_1B_{11} + L_2B_{22} + B_{33}$		
Stage 1	$\underbrace{u = -G_3\eta_3 + v}_{feedback}$	$\rightarrow$	$\begin{bmatrix} \dot{x}_I \\ \dot{\eta}_2 \\ \dot{\eta}_3 \end{bmatrix} = \begin{bmatrix} A_1 & A_{12}-A_{13}L_2 & A_{13}-B_{11}G_3 \\ 0 & A_2 & A_{23}+L_3A_{13}-B_2G_3 \\ 0 & 0 & A_3-B_3G_3 \end{bmatrix} \begin{bmatrix} x_I \\ \eta_2 \\ \eta_3 \end{bmatrix} + \begin{bmatrix} B_{11} \\ B_2 \\ B_3 \end{bmatrix} v$
Stage 2	$\underbrace{\xi_2 = \eta_2 - P_3\eta_3}_{transform}$	$\rightarrow$	$\begin{bmatrix} \dot{x}_I \\ \dot{\xi}_2 \\ \dot{\eta}_3 \end{bmatrix} = \begin{bmatrix} A_1 & A_{12}-A_{13}L_2 & A_{13}-B_{11}G_3+(A_{12}-A_{13}L_2)P_3 \\ 0 & A_2 & 0 \\ 0 & 0 & A_3-B_3G_3 \end{bmatrix} \begin{bmatrix} x_I \\ \xi_2 \\ \eta_3 \end{bmatrix} + \begin{bmatrix} B_{11} \\ B_2-P_3B_3 \\ B_3 \end{bmatrix} v$
	Solve for P_3		$A_2P_3 - P_3(A_3 - B_3G_3) + A_{23} + L_3A_{13} - B_2G_3 = 0$
	$\underbrace{v = -G_2\xi_2 + w}_{feedback}$		$\begin{bmatrix} \dot{x}_I \\ \dot{\xi}_2 \\ \dot{\eta}_3 \end{bmatrix} = \begin{bmatrix} A_1 & A_{12}-A_{13}L_2-B_{11}G_2 & A_{13}-B_{11}G_3+(A_{12}-A_{13}L_2)P_3 \\ 0 & A_2-(B_2-P_3B_3)G_2 & 0 \\ 0 & 0 & A_3-B_3G_3 \end{bmatrix} \begin{bmatrix} x_I \\ \xi_2 \\ \eta_3 \end{bmatrix} + \begin{bmatrix} B_{11} \\ B_2-P_3B_3 \\ B_3 \end{bmatrix} w$
Stage 3	Relabel matrix components from *Stage 2*	$\rightarrow$	$\begin{bmatrix} \dot{x}_I \\ \dot{\xi}_2 \\ \dot{\eta}_3 \end{bmatrix} = \begin{bmatrix} \alpha_{11} & \alpha_{12} & \alpha_{13} \\ 0 & \alpha_{22} & 0 \\ 0 & \alpha_{32} & \alpha_{33} \end{bmatrix} \begin{bmatrix} x_I \\ \xi_2 \\ \eta_3 \end{bmatrix} + \begin{bmatrix} \beta_1 \\ \beta_2 \\ \beta_2 \end{bmatrix} w$
	$\underbrace{\xi_1 = x_I - P_1\xi_2 - P_2\eta_3}_{transform}$	$\rightarrow$	$\begin{bmatrix} \dot{\xi}_1 \\ \dot{\xi}_2 \\ \dot{\eta}_3 \end{bmatrix} = \begin{bmatrix} \alpha_{11} & 0 & 0 \\ 0 & \alpha_{22} & 0 \\ 0 & \alpha_{32} & \alpha_{33} \end{bmatrix} \begin{bmatrix} \xi_1 \\ \xi_2 \\ \eta_3 \end{bmatrix} + \begin{bmatrix} \beta_1 - P_1\beta_2 - P_2\beta_3 \\ \beta_2 \\ \beta_2 \end{bmatrix} w$
	Solve forP_1,P_2		$\alpha_{11}P_1 - P_1\alpha_{22} - P_2\alpha_{32} + \alpha_{12} = 0, \alpha_{11}P_2 - P_2\alpha_{33} + \alpha_{13} = 0$
	$\underbrace{w = -G_1\xi_1}_{feedback}$	$\rightarrow$	$\begin{bmatrix} \dot{\xi}_1 \\ \dot{\xi}_2 \\ \dot{\eta}_3 \end{bmatrix} = \begin{bmatrix} \alpha_{11}-(\beta_1-P_1\beta_2-P_2\beta_3)G_1 & 0 & 0 \\ -\beta_2G_1 & \alpha_{22} & 0 \\ -\beta_3G_1 & \alpha_{32} & \alpha_{33} \end{bmatrix} \begin{bmatrix} \xi_1 \\ \xi_2 \\ \eta_3 \end{bmatrix}$
Transformation	$\begin{bmatrix} \xi_I \\ \xi_2 \\ \eta_3 \end{bmatrix} = \begin{bmatrix} I & -P_1 & -P_2 \\ 0 & I & 0 \\ 0 & 0 & I \end{bmatrix} \begin{bmatrix} x_I \\ \xi_2 \\ \eta_3 \end{bmatrix} = T_3 \begin{bmatrix} x_I \\ \xi_2 \\ \eta_3 \end{bmatrix} = T_3 \begin{bmatrix} I & 0 & 0 \\ 0 & I & -P_3 \\ 0 & 0 & I \end{bmatrix} \begin{bmatrix} x_I \\ \eta_2 \\ \eta_3 \end{bmatrix} = T_3T_2 \begin{bmatrix} I & 0 & 0 \\ L_3 & I & 0 \\ L_1 & L_2 & I \end{bmatrix} \begin{bmatrix} x_I \\ x_{II} \\ x_{III} \end{bmatrix} = T_3T_2T_1 \begin{bmatrix} x_I \\ x_{II} \\ x_{III} \end{bmatrix}$ $\boxed{\begin{bmatrix} \xi_I \\ \xi_2 \\ \eta_3 \end{bmatrix} = T_3T_2T_1 \begin{bmatrix} x_I \\ x_{II} \\ x_{III} \end{bmatrix} = T \begin{bmatrix} x_I \\ x_{II} \\ x_{III} \end{bmatrix} = \begin{bmatrix} I-P_1L_3-P_2L_1+P_1P_3L_1 & -P_1-P_2L_2+P_1P_3L_2 & P_1P_3-P_2 \\ L_3-P_3L_1 & I-P_3L_2 & -P_3 \\ L_1 & L_2 & I \end{bmatrix} \begin{bmatrix} x_I \\ x_{II} \\ x_{III} \end{bmatrix}}$		

Table 9.3 Three-stage continuous-time feedback controller design application to three-time scale linear control systems – introductory phase summary

Original partitioned system in *singularly perturbed form*	$\begin{bmatrix} \dot{x}_I \\ \varepsilon_1 \dot{x}_{II} \\ \varepsilon_2 \dot{x}_{III} \end{bmatrix} = \begin{bmatrix} A_{11} & A_{12} & A_{13} \\ A_{21} & A_{22} & A_{23} \\ A_{31} & A_{32} & A_{33} \end{bmatrix} \begin{bmatrix} x_I \\ x_{II} \\ x_{III} \end{bmatrix} + \begin{bmatrix} B_{11} \\ B_{22} \\ B_{33} \end{bmatrix} u = Ax + Bu,\ y = \begin{bmatrix} C_{11} & C_{22} & C_{33} \end{bmatrix} \begin{bmatrix} x_I \\ x_{II} \\ x_{III} \end{bmatrix}$
Upper triangular form	$\begin{bmatrix} \dot{x}_I \\ \varepsilon_1 \dot{\eta}_2 \\ \varepsilon_2 \dot{\eta}_3 \end{bmatrix} = \begin{bmatrix} A_1 & A_{12} - A_{13}L_2 & A_{13} \\ 0 & A_2 & A_{23} + \varepsilon_1 L_3 A_{13} \\ 0 & 0 & A_3 \end{bmatrix} \begin{bmatrix} x_I \\ \eta_2 \\ \eta_3 \end{bmatrix} + \begin{bmatrix} B_{11} \\ B_2 \\ B_3 \end{bmatrix} u$ $A_1 = (A_{11} - A_{13}L_1) - (A_{12} - A_{13}L_2)L_3$ $A_2 = A_{22} - A_{23}L_2 + L_3(A_{12} - A_{13}L_2),\ B_2 = B_{22} + L_3 B_{11}$ $A_3 = \varepsilon_2 L_1 A_{13} + \frac{\varepsilon_2}{\varepsilon_1} L_2 A_{23} + A_{33} B_3 = \varepsilon_2 L_1 B_{11} + \frac{\varepsilon_2}{\varepsilon_1} L_2 B_{22} + B_{33}$
Solve for L_1, L_2	$L_1^{(i+1)} = A_{33}^{-1}\left[A_{31} + \varepsilon_2 L_1^{(i)} A_{11} + \frac{\varepsilon_2}{\varepsilon_1} L_2^{(i)} A_{21} - \left(\varepsilon_2 L_1^{(i)} A_{13} + \frac{\varepsilon_2}{\varepsilon_1} L_2^{(i)} A_{23}\right) L_1^{(i)}\right],\ L_1^{(0)} = A_{33}^{-1} A_{31}$ $L_2^{(i+1)} = A_{33}^{-1}\left[A_{32} + \varepsilon_2 L_1^{(i)} A_{12} + \frac{\varepsilon_2}{\varepsilon_1} L_2^{(i)} A_{22} - \left(\varepsilon_2 L_1^{(i)} A_{13} + \frac{\varepsilon_2}{\varepsilon_1} L_2^{(i)} A_{23}\right) L_2^{(i)}\right],\ L_2^{(0)} = A_{33}^{-1} A_{32}$
Solve for L_3	$A_{22} - A_{23}L_2$ is invertible $L_3^{(i+1)} = (A_{22} - A_{23}L_2)^{-1}\left[A_{21} - A_{23}L_1 - \varepsilon_1 L_3^{(i)}(A_{12} - A_{13}L_2)L_3^{(i)} + \varepsilon_1 L_3^{(i)}(A_{11} - A_{13}L_1)\right]$ $L_3^{(0)} = (A_{22} - A_{23}L_2)^{-1}(A_{21} - A_{23}L_1)$ $A_{22} - A_{23}L_2$ is not invertible $(A_{22} - A_{23}L_2)L_3^{(i+1)} - \varepsilon_1 L_3^{(i+1)}(A_{11} - A_{13}L_1) = A_{21} - A_{23}L_1 - \varepsilon_1 L_3^{(i)}(A_{12} - A_{13}L_2)L_3^{(i)}$ Solve for $L_3^{(0)}$: $(A_{22} - A_{23}L_2)L_3^{(0)} - \varepsilon_1 L_3^{(0)}(A_{11} - A_{13}L_1) = A_{21} - A_{23}L_1$

Table 9.4 Three-stage continuous-time feedback controller design applications to three-time-scale linear control systems – three-stage procedure summary

Stage 1	Subsystem A_3 is already decoupled	
Stage 2	Solve for P_3	$P_3^{(i+1)} = (A_3 - B_3G_3)^{-1}\left(\dfrac{\varepsilon_2}{\varepsilon_1}\right)\left(A_2P_3^{(i)} + A_{23} + \varepsilon_1 L_3 A_{13} - B_2G_3\right),\ P_3^{(0)} = 0$
Stage 3	Solve forP_2,P_1	$P_2^{(i+1)} = \varepsilon_2\left(\alpha_{11}P_2^{(i)} + \alpha_{13}\right)\alpha_{33}^{-1},\ P_2^{(0)} = 0,\ i = 1,2\ldots N$ $P_1^{(i+1)} = \left[\varepsilon_1\left(\alpha_{12} + \alpha_{11}P_1^{(i)}\right) - \dfrac{\varepsilon_1}{\varepsilon_2}P_2^{(N)}\alpha_{32}\right]\alpha_{22}^{-1},\ P_1^{(0)} = \dfrac{\varepsilon_1}{\varepsilon_2}P_2^{(N)}\alpha_{32}\alpha_{22}^{-1}$

Appendix 9.2: Summary of the Four-Stage Continuous-Time Feedback Controller Design (Table 9.5 and 9.6)

Table 9.5 Four-stage feedback controller design – introductory phase summary

Original partitioned system	$\begin{bmatrix} \dot{x}_I \\ \dot{x}_{II} \\ \dot{x}_{III} \\ \dot{x}_{IV} \end{bmatrix} = \begin{bmatrix} A_{11} & A_{12} & A_{13} & A_{14} \\ A_{21} & A_{22} & A_{23} & A_{24} \\ A_{31} & A_{32} & A_{33} & A_{34} \\ A_{41} & A_{42} & A_{43} & A_{44} \end{bmatrix} \begin{bmatrix} x_I \\ x_{II} \\ x_{III} \\ x_{IV} \end{bmatrix} + \begin{bmatrix} B_{11} \\ B_{22} \\ B_{33} \\ B_{44} \end{bmatrix} u,\ y = \begin{bmatrix} C_{11} & C_{22} & C_{33} & C_{44} \end{bmatrix} \begin{bmatrix} x_I \\ x_{II} \\ x_{III} \\ x_{IV} \end{bmatrix}$
$\underbrace{\eta_4 = L_1 x_I + L_2 x_{II} + L_3 x_{III} + x_{IV}}_{transform}$	$\rightarrow \begin{bmatrix} \dot{x}_I \\ \dot{x}_{II} \\ \dot{x}_{III} \\ \dot{\eta}_{IV} \end{bmatrix} = \begin{bmatrix} a_{11} & a_{12} & a_{13} & a_{14} \\ a_{21} & a_{22} & a_{23} & a_{24} \\ a_{31} & a_{32} & a_{33} & a_{34} \\ f_{41}(L_1, L_2, L_3) & f_{42}(L_1, L_2, L_3) & f_{43}(L_1, L_2, L_3) & A_4 \end{bmatrix} \begin{bmatrix} x_I \\ x_{II} \\ x_{III} \\ \eta_{IV} \end{bmatrix} + \begin{bmatrix} b_{11} \\ b_{22} \\ b_{33} \\ B_4 \end{bmatrix} u$
	$a_{11} = A_{11} - A_{14}L_1$, $a_{12} = A_{12} - A_{14}L_2$, $a_{13} = A_{13} - A_{14}L_3$, $a_{14} = A_{14}$, $b_{11} = B_{11}$, $a_{21} = A_{21} - A_{24}L_1$, $a_{22} = A_{22} - A_{24}L_2$, $a_{23} = A_{23} - A_{24}L_3$, $a_{24} = A_{24}$, $b_{22} = B_{22}$, $a_{31} = A_{31} - A_{34}L_1$, $a_{32} = A_{32} - A_{34}L_2$, $a_{33} = A_{33} - A_{34}L_3$, $a_{34} = A_{34}$, $b_{33} = B_{33}$, $A_4 = A_{44} + L_1A_{14} + L_2A_{24} + L_3A_{34}$, $B_4 = B_{44} + L_1B_{11} + L_2B_{22} + L_3B_{33}$
Solve for L_1, L_2, L_3	$f_{41}(L_1, L_2, L_3) = L_1A_{11} + L_2A_{21} + L_3A_{31} + A_{41} - A_4L_1 = 0$ $f_{42}(L_1, L_2, L_3) = L_1A_{12} + L_2A_{22} + L_3A_{32} + A_{42} - A_4L_2 = 0$ $f_{43}(L_1, L_2, L_3) = L_1A_{13} + L_2A_{23} + L_3A_{33} + A_{43} - A_4L_3 = 0$
$\underbrace{\eta_3 = L_4 x_I + L_5 x_{II} + x_{III}}_{transform}$	$\rightarrow \begin{bmatrix} \dot{x}_I \\ \dot{x}_{II} \\ \dot{\eta}_3 \\ \dot{\eta}_4 \end{bmatrix} = \begin{bmatrix} \alpha_{11} & \alpha_{12} & \alpha_{13} & \alpha_{14} \\ \alpha_{21} & \alpha_{22} & \alpha_{23} & \alpha_{24} \\ f_{31}(L_4, L_5) & f_{32}(L_4, L_5) & a_{33} + L_4a_{13} + L_5a_{23} & a_{34} + L_4a_4 + L_5a_{24} \\ 0 & 0 & 0 & A_4 \end{bmatrix} \begin{bmatrix} x_I \\ x_{II} \\ \eta_3 \\ \eta_4 \end{bmatrix} + \begin{bmatrix} \beta_{11} \\ \beta_{22} \\ b_{33} + L_4b_{11} + L_5b_{22} \\ B_4 \end{bmatrix} u$ $\alpha_{11} = a_{11} - a_{13}L_4, \alpha_{12} = a_{12} - a_{13}L_5, \alpha_{13} = a_{13}, \alpha_{14} = a_{14}, \beta_{11} = b_{11}\alpha_{21} = a_{21} - a_{23}L_4, \alpha_{22} = a_{22} - a_{23}L_5, \alpha_{23} = a_{23}, \alpha_{24} = a_{24}, \beta_{22} = b_{22}$
Solve for L_4, L_5	$f_{31}(L_4, L_5) = (a_{31} + L_4a_{11} + L_5a_{21}) - (a_{33} + L_4a_{13} + L_5a_{23})L_4 = 0$ $f_{32}(L_4, L_5) = (a_{32} + L_4a_{12} + L_5a_{22}) - (a_{33} + L_4a_{13} + L_5a_{23})L_5 = 0$
$\underbrace{\eta_2 = L_6 x_I + x_{II}}_{transform}$	$\rightarrow \begin{bmatrix} \dot{x}_I \\ \dot{\eta}_2 \\ \dot{\eta}_3 \\ \dot{\eta}_4 \end{bmatrix} = \begin{bmatrix} \alpha_{11} - \alpha_{12}L_6 & \alpha_{12} & \alpha_{13} & a_{14} \\ f_{21}(L_6) & \alpha_{22} + L_6\alpha_{12} & \alpha_{23} + L_6\alpha_{13} & \alpha_{24} + L_6\alpha_{14} \\ 0 & 0 & \alpha_{33} & \alpha_{34} \\ 0 & 0 & 0 & A_{44} \end{bmatrix} \begin{bmatrix} x_I \\ \eta_2 \\ \eta_3 \\ \eta_4 \end{bmatrix} + \begin{bmatrix} \beta_{11} \\ \beta_{22} \\ \beta_{33} \\ B_4 \end{bmatrix} u = \bar{A} \begin{bmatrix} x_I \\ \eta_2 \\ \eta_3 \\ \eta_4 \end{bmatrix} + \bar{B}u$

(continued)

Table 9.5 (continued)

Solve forL_6	$f_{21}(L_6) = \alpha_{21}+L_6\alpha_{11} - (\alpha_{22}+L_6\alpha_{12})L_6 = 0$
Similarity transform	$\begin{bmatrix} x_I \\ \eta_2 \\ \eta_3 \\ \eta_4 \end{bmatrix} = \begin{bmatrix} I & 0 & 0 & 0 \\ L_6 & I & 0 & 0 \\ L_4 & L_5 & I & 0 \\ L_1 & L_2 & L_3 & I \end{bmatrix} \begin{bmatrix} x_I \\ x_{II} \\ x_{III} \\ x_{IV} \end{bmatrix} = T_1 \begin{bmatrix} x_I \\ x_{II} \\ x_{III} \\ x_{IV} \end{bmatrix}, \begin{bmatrix} x_I \\ x_{II} \\ x_{III} \\ x_{IV} \end{bmatrix} = \begin{bmatrix} I & 0 & 0 & 0 \\ -L_6 & I & 0 & 0 \\ L_5L_6 - L_4 & -L_5 & I & 0 \\ L_3L_4 + L_2L_6 - L_1 & L_3L_5 - L_2 & -L_3 & I \end{bmatrix} \begin{bmatrix} x_I \\ \eta_2 \\ \eta_3 \\ \eta_4 \end{bmatrix} = T_1^{-1} \begin{bmatrix} x_I \\ \eta_2 \\ \eta_3 \\ \eta_4 \end{bmatrix}$

Table 9.6 Four-stage feedback controller design – four-stage procedure summary

Introduce notation	$\begin{bmatrix}\dot{x}_I\\ \dot{\eta}_2\\ \dot{\eta}_3\\ \dot{\eta}_4\end{bmatrix}=\begin{bmatrix}\bar{A}_{11}&\bar{A}_{12}&\bar{A}_{13}&\bar{A}_{14}\\0&\bar{A}_{22}&\bar{A}_{23}&\bar{A}_{24}\\0&0&\bar{A}_{33}&\bar{A}_{34}\\0&0&0&\bar{A}_{44}\end{bmatrix}\begin{bmatrix}x_I\\ \eta_2\\ \eta_3\\ \eta_4\end{bmatrix}+\begin{bmatrix}\bar{B}_{11}\\ \bar{B}_{22}\\ \bar{B}_{33}\\ \bar{B}_{44}\end{bmatrix}u$		$\bar{A}_{11}=\alpha_{11}-\alpha_{12}L_6,\bar{A}_{12}=\alpha_{12},\bar{A}_{13}=\alpha_{13},\bar{A}_{14}=\alpha_{14}$ $\bar{A}_{22}=\alpha_{22}+L_6\alpha_{12},\bar{A}_{23}=\alpha_{23}+L_6\alpha_{13},\bar{A}_{24}=\alpha_{24}+L_6\alpha_{14}$ $\bar{A}_{33}=\alpha_{33},\bar{A}_{34}=\alpha_{34},A_{44}=A_4=a_{44}$ $\bar{B}_{11}=\beta_{11},\bar{B}_{22}=B_{22},\bar{B}_{33}=\beta_{33},\bar{B}_{44}=B_{44}$
Stage 1	$\underbrace{u=-G_4\eta_4+v}_{feedback}$	$\rightarrow$	$\begin{bmatrix}\dot{x}_I\\ \dot{\eta}_2\\ \dot{\eta}_3\\ \dot{\eta}_4\end{bmatrix}=\bar{A}\begin{bmatrix}x_I\\ \eta_2\\ \eta_3\\ \eta_4\end{bmatrix}+\bar{B}u=\begin{bmatrix}\bar{A}_{11}&\bar{A}_{12}&\bar{A}_{13}&\bar{A}_{14}-\bar{B}_{11}G_4\\0&\bar{A}_{22}&\bar{A}_{23}&\bar{A}_{24}-\bar{B}_{22}G_4\\0&0&\bar{A}_{33}&\bar{A}_{34}-\bar{B}_{33}G_4\\0&0&0&\bar{A}_{44}-\bar{B}_{44}G_4\end{bmatrix}\begin{bmatrix}x_I\\ \eta_2\\ \eta_3\\ \eta_4\end{bmatrix}+\begin{bmatrix}\bar{B}_{11}\\ \bar{B}_{22}\\ \bar{B}_{33}\\ \bar{B}_{44}\end{bmatrix}v$
Stage 2	$\underbrace{\xi_3=\eta_3-P_3\eta_4}_{transform}$	$\rightarrow$	$\begin{bmatrix}\dot{x}_I\\ \dot{\eta}_2\\ \dot{\eta}_3\\ \dot{\eta}_4\end{bmatrix}=\begin{bmatrix}\bar{A}_{11}&\bar{A}_{12}&\bar{A}_{13}&\bar{A}_{14}-\bar{B}_{11}G_4+\bar{A}_{13}P_3\\0&\bar{A}_{22}&\bar{A}_{23}&\bar{A}_{24}-\bar{B}_{22}G_4+\bar{A}_{23}P_3\\0&0&\bar{A}_{33}&0\\0&0&0&\bar{A}_{44}-\bar{B}_{44}G_4\end{bmatrix}\begin{bmatrix}x_I\\ \eta_2\\ \eta_3\\ \eta_4\end{bmatrix}+\begin{bmatrix}\bar{B}_{11}\\ \bar{B}_{22}\\ \bar{B}_{33}-P_3\bar{B}_{44}\\ \bar{B}_{44}\end{bmatrix}v$
	Solve forP_3		$\bar{A}_{33}P_3-P_3(\bar{A}_{44}-\bar{B}_{44}G_4)+\bar{A}_{33}-\bar{B}_{33}G_4=0$
	$\underbrace{v=-G_3\xi_3+w}_{feedback}$	$\rightarrow\begin{bmatrix}\dot{x}_I\\ \dot{\eta}_2\\ \dot{\eta}_3\\ \dot{\eta}_4\end{bmatrix}=$	$\begin{bmatrix}\bar{A}_{11}&\bar{A}_{12}&\bar{A}_{13}-\bar{B}_{11}G_3&\bar{A}_{14}-\bar{B}_{11}G_4+\bar{A}_{13}P_3\\0&\bar{A}_{22}&\bar{A}_{23}-\bar{B}_{22}G_3&\bar{A}_{24}-\bar{B}_{22}G_4+\bar{A}_{23}P_3\\0&0&\bar{A}_{33}-(\bar{B}_{33}-P_3\bar{B}_{44})G_3&0\\0&0&-\bar{B}_{44}G_3&\bar{A}_{44}-\bar{B}_{44}G_4\end{bmatrix}\begin{bmatrix}x_I\\ \eta_2\\ \eta_3\\ \eta_4\end{bmatrix}+\begin{bmatrix}\bar{B}_{11}\\ \bar{B}_{22}\\ \bar{B}_{33}-P_3\bar{B}_{44}\\ \bar{B}_{44}\end{bmatrix}v$
Stage 3	$\underbrace{\xi_2=\eta_2-\;-P_{23}\xi_3-\;-P_{24}\eta_4}_{transform}$	$\rightarrow$	$\begin{bmatrix}\dot{x}_I\\ \dot{\xi}_2\\ \dot{\xi}_3\\ \dot{\eta}_4\end{bmatrix}=\begin{bmatrix}\bar{A}_{11}&\bar{A}_{12}&\bar{A}_{13}-\bar{B}_{11}G_3+\bar{A}_{12}P_{23}&S_{14}\\0&\bar{A}_{22}&0&0\\0&0&\bar{A}_{33}-(\bar{B}_{33}-P_3\bar{B}_{44})G_3&0\\0&0&-\bar{B}_{44}G_3&\bar{A}_{44}-\bar{B}_{44}G_4\end{bmatrix}\begin{bmatrix}x_I\\ \xi_2\\ \xi_3\\ \eta_4\end{bmatrix}+\begin{bmatrix}\bar{B}_{11}\\ \bar{B}_{22}+\bar{B}_{33}-P_3\bar{B}_{44}-P_{24}\bar{B}_{44}\\ \bar{B}_{33}-P_3\bar{B}_{44}\\ \bar{B}_{44}\end{bmatrix}wS_{14}=\bar{A}_{14}-\bar{B}_{11}G_4+\bar{A}_{13}P_3+\bar{A}_{12}P_{24}$
	Solve for P_{23},P_{24}		$\bar{A}_{22}P_{23}-P_{23}\left(\bar{A}_{33}-(\bar{B}_{33}-P_3\bar{B}_{44})G_3\right)+P_{24}\bar{B}_{44}G_3+\bar{A}_{23}-\bar{B}_{22}G_3=0$ $\bar{A}_{22}P_{24}-P_{24}(\bar{A}_{44}-\bar{B}_{44}G_4)+\bar{A}_{24}-\bar{B}_{22}G_4+\bar{A}_{23}P_3=0$
	$\underbrace{w=-G_2\xi_2+f}_{feedback}$	$\rightarrow$	$\begin{bmatrix}\dot{x}_I\\ \dot{\xi}_2\\ \dot{\xi}_3\\ \dot{\eta}_4\end{bmatrix}=\begin{bmatrix}\bar{A}_{11}&\bar{A}_{12}-\bar{B}_{11}G_2&\bar{A}_{13}-\bar{B}_{11}G_3+\bar{A}_{12}P_{23}&S_{14}\\0&\bar{A}_{22}-\bar{B}_{2\xi}G_2&0&0\\0&-\bar{B}_{3\xi}G_2&\bar{A}_{33}-(\bar{B}_{33}-P_3\bar{B}_{44})G_3&0\\0&-\bar{B}_{44}G_2&-\bar{B}_{44}G_3&\bar{A}_{44}-\bar{B}_{44}G_4\end{bmatrix}\begin{bmatrix}x_I\\ \xi_2\\ \xi_3\\ \eta_4\end{bmatrix}+\begin{bmatrix}\bar{B}_{11}\\ \bar{B}_{2\xi}\\ \bar{B}_{3\xi}\\ \bar{B}_{44}\end{bmatrix}f$ $\bar{B}_{2\xi}=\bar{B}_{22}+P_{23}(\bar{B}_{33}-P_3\bar{B}_{44})-P_{23}\bar{B}_{44},\ \bar{B}_{3\xi}=\bar{B}_{33}-P_3\bar{B}_{44}$
	Introduce notation		$\begin{bmatrix}\dot{x}_I\\ \dot{\xi}_2\\ \dot{\xi}_3\\ \dot{\eta}_4\end{bmatrix}=\begin{bmatrix}S_{11}&S_{12}&S_{13}&S_{14}\\0&S_{22}&0&0\\0&S_{23}&S_{33}&0\\0&S_{42}&S_{43}&S_{44}\end{bmatrix}\begin{bmatrix}x_I\\ \xi_2\\ \xi_3\\ \eta_4\end{bmatrix}+\begin{bmatrix}Q_1\\ Q_2\\ Q_3\\ Q_4\end{bmatrix}f$ $S_{11}=\bar{A}_{11},S_{12}=\bar{A}_{12}-\bar{B}_{11}G_2,S_{13}=\bar{A}_{13}-\bar{B}_{11}G_3+\bar{A}_{12}P_{23},S_{22}=\bar{A}_{22}-\bar{B}_{2\xi}G_2,$ $S_{32}=-\bar{B}_{3\xi}G_2,S_{33}=\bar{A}_{33}-(\bar{B}_{33}-P_3\bar{B}_{44})G_3,S_{42}=-\bar{B}_{44}G_2,S_{43}=-\bar{B}_{44}G_3,$ $S_{44}=\bar{A}_{44}-\bar{B}_{44}G_4,Q_1=\bar{B}_{11},Q_2=\bar{B}_{2\xi},Q_3=\bar{B}_{3\xi},Q_4=\bar{B}_{44}$

(continued)

Table 9.6 (continued)

Stage 4	$\underbrace{\xi_1 = x_I - P_{12}\xi_2 - P_{13}\xi_3 - P_{14}\eta_4}_{transform}$	→	$\begin{bmatrix}\dot{\xi}_1\\ \dot{\xi}_2\\ \dot{\xi}_3\\ \dot{\eta}_4\end{bmatrix} = \begin{bmatrix}S_{11} & 0 & 0 & 0\\ 0 & S_{22} & 0 & 0\\ 0 & S_{23} & S_{33} & 0\\ 0 & S_{42} & S_{43} & S_{44}\end{bmatrix}\begin{bmatrix}\xi_1\\ \xi_2\\ \xi_3\\ \eta_4\end{bmatrix} + \begin{bmatrix}Q_1 - P_{12}Q_2 - P_{13}Q_3 - P_{14}Q_4\\ Q_2\\ Q_3\\ Q_4\end{bmatrix} f$
	Solve for P_{12}, P_{13}, P_{14}		$S_{11}P_{12} - P_{12}S_{22} - P_{13}S_{32} - P_{14}S_{42} + S_{12} = 0$ $S_{11}P_{13} - P_{13}S_{33} - P_{14}S_{43} + S_{13} = 0,\ S_{11}P_{14} - P_{14}S_{44} + S_{14} = 0$
	$\underbrace{f = -G_1\xi_1}_{feedback}$	→	$\begin{bmatrix}\dot{\xi}_1\\ \dot{\xi}_2\\ \dot{\xi}_3\\ \dot{\eta}_4\end{bmatrix} = \begin{bmatrix}S_{11} - (Q_1 - P_{12}Q_2 - P_{13}Q_3 - P_{14}Q_4)G_1 & 0 & 0 & 0\\ -Q_2G_1 & S_{22} & 0 & 0\\ -Q_3G_1 & S_{23} & S_{33} & 0\\ -Q_4G_1 & S_{42} & S_{43} & S_{44}\end{bmatrix}\begin{bmatrix}\xi_1\\ \xi_2\\ \xi_3\\ \eta_4\end{bmatrix}$
Transformation	$\begin{bmatrix}x_I\\ \eta_2\\ \xi_3\\ \eta_4\end{bmatrix} = \begin{bmatrix}I & 0 & 0 & 0\\ 0 & I & 0 & 0\\ 0 & 0 & I & -P_3\\ 0 & 0 & 0 & I\end{bmatrix}\begin{bmatrix}x_I\\ \eta_2\\ \eta_3\\ \eta_4\end{bmatrix} = T_2\begin{bmatrix}x_I\\ \eta_2\\ \eta_3\\ \eta_4\end{bmatrix},\ \begin{bmatrix}x_I\\ \eta_2\\ \eta_3\\ \eta_4\end{bmatrix} = \begin{bmatrix}I & 0 & 0 & 0\\ 0 & I & 0 & 0\\ 0 & 0 & I & P_3\\ 0 & 0 & 0 & I\end{bmatrix}\begin{bmatrix}x_I\\ \eta_2\\ \xi_3\\ \eta_4\end{bmatrix} = T_2^{-1}\begin{bmatrix}x_I\\ \eta_2\\ \xi_3\\ \eta_4\end{bmatrix},\ \begin{bmatrix}x_I\\ \xi_2\\ \xi_3\\ \eta_4\end{bmatrix} = \begin{bmatrix}I & 0 & 0 & 0\\ 0 & I & -P_{23} & -P_{24}\\ 0 & 0 & I & 0\\ 0 & 0 & 0 & I\end{bmatrix}\begin{bmatrix}x_I\\ \eta_2\\ \xi_3\\ \eta_4\end{bmatrix} = T_3\begin{bmatrix}x_I\\ \eta_2\\ \xi_3\\ \eta_4\end{bmatrix},\ \begin{bmatrix}x_I\\ \eta_2\\ \xi_3\\ \eta_4\end{bmatrix} = \begin{bmatrix}I & 0 & 0 & 0\\ 0 & I & P_{23} & P_{24}\\ 0 & 0 & I & 0\\ 0 & 0 & 0 & I\end{bmatrix}\begin{bmatrix}x_I\\ \xi_2\\ \xi_3\\ \eta_4\end{bmatrix} = T_3^{-1}\begin{bmatrix}x_I\\ \xi_2\\ \xi_3\\ \eta_4\end{bmatrix}$ $\begin{bmatrix}\xi_1\\ \xi_2\\ \xi_3\\ \eta_4\end{bmatrix} = \begin{bmatrix}I & -P_{12} & -P_{13} & -P_{14}\\ 0 & I & 0 & 0\\ 0 & 0 & I & 0\\ 0 & 0 & 0 & I\end{bmatrix}\begin{bmatrix}x_I\\ \xi_2\\ \xi_3\\ \eta_4\end{bmatrix} = T_4\begin{bmatrix}x_I\\ \xi_2\\ \xi_3\\ \eta_4\end{bmatrix},\ \begin{bmatrix}x_I\\ \xi_2\\ \xi_3\\ \eta_4\end{bmatrix} = \begin{bmatrix}I & P_{12} & P_{13} & P_{14}\\ 0 & I & 0 & 0\\ 0 & 0 & I & 0\\ 0 & 0 & 0 & I\end{bmatrix}\begin{bmatrix}\xi_1\\ \xi_2\\ \xi_3\\ \eta_4\end{bmatrix} = T_4^{-1}\begin{bmatrix}\xi_1\\ \xi_2\\ \xi_3\\ \eta_4\end{bmatrix}$ $\boxed{\begin{bmatrix}x_I\\ x_{II}\\ x_{III}\\ x_{IV}\end{bmatrix} = T_1^{-1}\begin{bmatrix}x_I\\ \eta_2\\ \eta_3\\ \eta_4\end{bmatrix} = T_1^{-1}T_2^{-1}\begin{bmatrix}x_I\\ \eta_2\\ \xi_3\\ \eta_4\end{bmatrix} = T_1^{-1}T_2^{-1}T_3^{-1}\begin{bmatrix}\eta_1\\ \xi_2\\ \xi_3\\ \eta_4\end{bmatrix} = T_1^{-1}T_2^{-1}T_3^{-1}T_4^{-1}\begin{bmatrix}\xi_1\\ \xi_2\\ \xi_3\\ \eta_4\end{bmatrix},\ \begin{bmatrix}\xi_1\\ \xi_2\\ \xi_3\\ \eta_4\end{bmatrix} = T_4T_3T_2T_1\begin{bmatrix}x_I\\ x_{II}\\ x_{III}\\ x_{IV}\end{bmatrix} = T\begin{bmatrix}x_I\\ x_{II}\\ x_{III}\\ x_{IV}\end{bmatrix}}$

References

Abdin Z, Webb C, Mac E, Gray A (2017) PEM fuel cell model and simulation in MATLAB-Simulink based on physical parameters. Energy 116:1131–1144
Adachi H, Ahmet S, Lee S, Papadis D, Ahluwalla R, Bendert J, Kanner S, Yamazaki Y (2009) A natural-gas fuel processor for a residential fuel cell system. J Power Sources 188:244–255
Ali J, Hoang N, Hissain M, Dochain D (2015) Review and classification of recent observers applied in chemical process systems. Comput Chem Eng 76:27–41
Amjadifard R, Beheshti M, Yazdanpaanah M (2011) Robust stabilization for a singularly perturbed systems. Trans ASME J Dyn Syst Meas Control 133:051004-1–051004-6
Arsov G (2007) Parametric PSPICE model of a PEM fuel cell. Electronics 11:99–103
Barbir F (2005) PEM fuel cells: theory and practice. Elsevier, Amsterdam
Barelli L, Bidini G, Gallorini F, Ottaviano A (2012) Dynamic analysis of PEMFC-based CHP systems for domestic application. Appl Energy 91:13–28
Baruh H, Choe K (1990) Sensor placement in structural control. J Guid Dyn Control 13:524–533
Barzegari M, Dardel M, Alizadeh E, Ramiar A (2016) Reduced-order model of cascade type PEM fuel cell stack with integrated humidifiers and water separators. Energy 113:683–692
Bavarian M, Soroush M, Kevredkidis I, Benziger J (2010) Mathematical modeling, steady state and dynamic behavior, and control of fuel cells: a review. Ind Eng Chem Res 49:7922–7950
Becherif M, Hissel D, Gaagat S, Wack M (2011) Electrical equivalent model of a proton exchange membrane fuel cell with experimental validation. Renew Energy 36:2582–2588
Benziger J, Satterfield M, Hogarth W, Nehlsen J, Kevrekidis I (2006) The power performance curve for engineering analysis of fuel cells. J Power Sources 155:272–285
Bhargav A, Lyubovsky M, Dixit M (2014) Managing fuel variabality in LPG-based PEM fuel cell systems: — I: theormodinamic simulations. Int J Hydrogen Energy 39:17231–17239
Bidani M, Randhy N, Bensassi B (2002) Optimal control of discrete-time singularly perturbed systems. Int J Control 75:955–966
Bingulac S, Van Landingham H (1993) Algorithms for computer aided design of multivariable control systems. Marcel Dekker, New York
Chakraborty U (2018) Reversible and irreversible potentials and an inaccuracy in popular models in the fuel cell literature. Energies 11:1851. https://doi.org/10.3390/en11071851
Chen T (2012) Linear system theory and design. Oxford University Press, Oxford, UK
Chen P-C (2013) Robust voltage tracking control for proton exchange membrane fuel cells. Energ Conver Manage 65:408–419

V. Radisavljević-Gajić et al., *Multi-Stage and Multi-Time Scale Feedback Control of Linear Systems with Applications to Fuel Cells*, Mechanical Engineering Series,
https://doi.org/10.1007/978-3-030-10389-7

Chen C-F, Pan S-T, Hsieh J-G (2002) Stability analysis of a class of uncertain discrete singularly perturbed systems with multiple time delays. Trans ASME J Dyn Syst Meas Control 124:467–472

Chiu L, Diong B, Gemmen R (2004) An improved small signal model of the dynamic behavior of PEM fuel cells. IEEE Trans Ind Appl 40:970–977

Cipiti F, Pino L, Vita A, Lagana M, Rucupero V (2013) Experimental investigation on a methane fuel processor for polymer electrolyte fuel cells. Int J Hydrogen Energy 38:2387–2397

Cronin J (2008) Mathematical aspects of Hodgkin-Huxley neural theory. Cambridge University Press, Cambridge

Daud W, Rosli R, Majlan E, Hamid S, Mohamed R, Husaini T (2017) PEM fuel cell system control: a review. Renew Energy 113:620–638

Demetriou M, Kazantzis N (2005) Natural observer design for singularly perturbed vector second-order systems. Trans ASME J Dyn Syst Meas Control 127:648–655

Dimitriev M, Kurina G (2006) Singular perturbations in control systems. Autom Remote Control 67:1–43

Dong X, El-Gorashi T, Elmirghani J (2012) Use of renewable energy in an IP over WDM network with data centers. IET Optoelectron 6:155–164

Eikerling M, Kulikovsky A (2014) Polymer electrolyte fuel cells: physical principles of materials and operation. CRC Press, Boca Raton

El-Sharkh M, Rahman A, Alam M, Byrne P, Sakla A, Thomas T (2004) A dynamic model for stand-alone PEM fuel cell power plant for residential applications. J Power Sources 138:199–204

El-Sharkh M, Sisworahardjo N, Uzunoglu M, Onar O, Alam M (2007) Dynamic behavior of PEM fuel cell and microturbine power plants. J Power Sources 164:315–321

Esteban S, Gordillo F, Aracil J (2013) Three-time scale singular perturbation control and stability analysis for an autonomous helicopter on a platform. Int J Robust Nonlinear Control 23:1360–1392

Famouri P, Gemmen R (2003) Electromechanical circuit model of a PEM fuel cell. In: Proceedings of Power Engineering Society regular meeting, pp 1436–1440

Franklin G, Powel J, Workman M (1990) Digital control of dynamic systems. Addison Wesley, Reading

Fuhrmann J, Haasdonk B, Holzbecher E, Ohlberger M (2008) Modeling and simulation of PEM fuel cells. ASME J Fuel Cell Sci Technol 5:020301-1

Gajic Z (2003) Linear dynamic systems and signals. Prentice Hall, Upper Saddle River

Gajic Z, Lelic M (1996) Modern control systems engineering. Prentice Hall International, London

Gajic Z, Lim M-T (2001) Optimal control of singularly perturbed linear systems and applications. Marcel Dekker, New York

Gajic Z, Qureshi M (1995) Matrix Lyapunov equation in system stability and control. Academic Press, San Diego

Gajic Z, Lim M-T, Skataric D, Su W-C, Kecman V (2009) Optimal control: weakly coupled systems and applications. CRC Press Taylor & Francis Group, Boca Raton

Gao Y-H, Bai Z-Z (2010) On inexact Newton methods based on doubling iteration scheme for non-symmetric algebraic Riccati equations. Numer Linear Algebra Appl. https://doi.org/10.1002/nla.727

Gawronski W (1994) A balanced LQG compensator for flexible structures. Automatica 30:1555–1564

Gawronski W (1998) Dynamics and control of structures: a modal approach. Springer, New York

Gawronski W, Juang J-N (1990) Model reduction for flexible structures. Control Dyn Syst 36:143–222

Gemmen R (2003) Analysis for the effect of inverter ripple current on fuel cell operating condition. J Fluid Eng 124:576–585

Golub G, Van Loan C (2013) Matrix computations, 4th edition. The Johns Hopkins University Press, Baltimore, MD, USA

Gou B, Na W, Diong B (2010) Fuel cells: modeling, control, and applications. CRC Press Taylor & Francis Group, Boca Raton

Graham R, Knuth D, Patashnik O (1989) Concrete mathematics. Addison-Wesley, Reading

Grujicic M, Chttajallu K, Law E, Pukrushpan J (2004a) Model-based control strategies in the dynamic interaction of air supply and fuel cell. Proc Inst Mech Eng A J Power Energy 218:487–499

Grujicic M, Chttajallu KM, Pukrushpan JT (2004b) Control of the transient behavior of polimer electrolyte membrane fuel cell systems. Proc Inst Mech Eng D J Automot Eng 218:1239–1250

Haddad A, Mannah M, Bazzi H (2015) Nonlinear time-variant model of the PEM type fuel cell for automotive applications. Simul Model Pract Theory 51:31–44

Hajizadeh A, Golkar M (2010) Intelligent robust control of hybrid distributed generation system under voltage sag. Expert Syst Appl 37:7627–7638

Han J, Yu S, Yi S (2017) Advanced thermal management automotive fuel cells using a model reference adaptive control algorithm. Int J Hydrogen Energy 42:4328–4341

Hayati M, Khayatian A, Dehghani M (2016) Simultaneous optimization of net power and enhancement of PEM fuel cell lifespan using extremum seeking and sliding mode control techniques. IEEE Trans Energy Convers 32:688–696

Headley A, Yu V, Borduin R, Chen D, Li W (2016) Development and experimental validation of a physics-based PEM fuel cell model for cathode humidity control design. IEEE/ASME Trans Mechatron 21:1778–1782

Hoffmann P, Dorgan B (2012) Tomorrow's energy: hydrogen, 'fuel cells, and prospects for a cleaner planet. MIT Press, Cambridge, MA

Hong L, Chen J, Liu Z, Huang L, Wu Z (2017) A nonlinear control strategy for fuel cell delivery in PEM fuel cells considering nitrogen permeation. Int J Hydrogen Energy 42:1565–1576

Hsiao FH, Hwang JD, ST P (2001) Stabilization of discrete singularly perturbed systems under composite observer-based controller. Trans ASME J Dyn Syst Meas Control 123:132–139

Jalics J, Krupa M, Rotstein H (2010) Mixed-mode oscillations in a three time-scale system of ODEs motivated by a neuronal model. Dyn Syst Int J 25:445–482

Jiao J (2014) Maximum power point tracking of fuel cell power system using fuzzy logic control. Electrotehn Electron Automat 62:45–52

Johnson CD (1988) Optimal initial conditions for full-order observers. Int J Control 48:857–864

Kalman R (1960) Contributions to the theory of optimal control. Bol Soc Mat Mexicana 5:102–119

Kalman R (1963) Mathematical description of linear dynamical systems. SIAM J Control 1:152–192

Karstedt J, Ogrzewalla J, Severin C, Pischinger S (2011) Development and design of experiments optimization of a high temperature proton exchange membrane fuel cell auxiliary power unit with onboard fuel processor. J Power Sources 196:9998–10009

Khalil H (2002) Nonlinear systems. Prentice Hall, Upper Saddle River

Kim B-S, Kim Y-J, Lim M-T (2004) LQG control for nonstandard singularly perturbed discrete-time systems. Trans ASME J Dyn Syst Meas Control 126:860–864

Kirk D (2004) Optimal control theory: an introduction. Dover Publications, Mineola

Kokotovic P, Khalil H, O'Reilly J (1999) Singular perturbation methods in control: analysis and design. Academic Press, Orlando

Kuehn C (2015) Multiple time scale dynamics. Springer, Cham

Kulikovsky A (2010) Analytical modeling of fuel cells. Elsevier, Amsterdam

Kummrow A, Emde M, Baltsuka A, Pshebichnikov M, Wiersma D (1999) Z Phys Chem Int J Res Phys Chem Chem Phys 212:153–159

Kunusch C, Mayosky M, Husar A (2011) Control-oriented modeling and experimental validation of a PEMFC generation system. IEEE Trans Energy Convers 26:851–861

Laghrouche S, Harmouche M, Ahmed F, Chitour Y (2015) Control of PEMFC air-feed system using Lyapunov-based robust and adaptive higher order sliding mode control. IEEE Trans Control Syst Technol 23:1594–1601

Larminie J, Dicks A (2001) Fuel cell systems explained. Wiley, New York

Laurim D, Salcedo J, Garcia-Nieto S, Martinez M (2010) Model predictive control relevant identification: multiple input multiple output against multiple input single output. IET Control Theory Appl 4:1756–1766

Lee C, Othmer G (2010) A multi-time-scale analysis of chemical reaction networks: I. Deterministic systems. J Math Biol 60:387–450

Lee D, Lee H, Lee K, Kim S (2007) A compact and highly efficient natural gas fuel processor for 1-kW residential polymer electrolyte membrane fuel cells. J Power Sources 165:337–341

Li Y, Rajakaruna S (2005) An analysis of the control and operation of a solid oxide fuel-cell power plant in an isolated system. IEEE Trans Energy Convers 20:381–387

Li D, Li C, Gao Z, Jin Q (2015a) On active disturbance rejection of the proton exchange membrane fuel cells. J Power Sources 283:452–463

Li Y, Zhao X, Tao S, Li Q, Chen W (2015b) Experimental study on anode and cathode pressure difference control and effects in a proton exchange membrane fuel cell system. Energ Technol 3:946–954

Litkouhi B, Khalil H (1984) Infinite-time regulators for singularly perturbed difference equations. Int J Control 39:587–598

Litkouhi B, Khalil H (1985) Multirate and composite control of two-time-scale discrete-time systems. IEEE Trans Autom Control 30:645–651

Liu H, Sun F, He K (2003) Survey of singularly perturbed control systems: theory and applications. Control Theory Appl 20:1–7

Mahmoud M (1986) Stabilization of discrete systems with multiple time scales. IEEE Trans Autom Control 31:159–162

Majlan E, Rohendi D, Daud W, Husaini T, Haque M (2018) Electrode for proton exchange membrane fuel cells: a review. Renew Sustain Energy Rev 89:117–134

Matraji I, Laghrouche S, Wack M (2012) Pressure control of a PEM fuel cell via second order sliding mode. Int J Hydrogen Energy 37:16104–16116

Matraji I, Laghtouche S, Jemei S, Wack M (2013) Robust control of the PEM fuel cell air-feed system via sub-optimal second order sliding model. Appl Energy 104:945–957

Matraji I, Ahmed F, Laghrouche S, Wack M (2015) Comparison of robust and adaptive second order sliding mode control in PEMFC air-feed systems. Int J Hydrogen Energy 40:9491–9504

McCain B, Stefanopoulou A, Siegel J (2010) Controllability and observability analysis of the liquid water distribution inside the gas diffusion layer of a unit fuel cell model. Trans ASME J Dyn Syst Meas Control 132:061303-1–051303-8

Medanic J (1982) Geometric properties and invariant manifolds of the Riccati equation. IEEE Trans Autom Control 27:670–677

Meirovich L, Baruh H (1983) On the problem of observation spillover in self-adjoint distributed parameter systems. J Optim Theory Appl 39:269–291

Milanovic M, Radisavljevic-Gajic V (2018) Optimal linear-quadratic integral feedback controller design with disturbance rejection for a proton exchange membrane fuel cell. In: ASME dynamic systems and control conference, Atlanta

Milanovic M, Rose P, Radisavljevic-Gajic V, Clayton G (2017) Five state analytical model proton exchange membrane fuel cell. In: ASME dynamic systems and control conference, Tysons Corner

Min K, Kang S, Mueller F, Auckland J, Brouwer J (2009) Dynamic simulation of a stationary proton exchange membrane fuel cell system. ASME J Fuel Cell Sci Technol 6:041015-1

Mitchell W, Bowers B, Garnier C, Boudjema F (2006) Dynamic behavior of gasoline fuel cell electric vehicles. J Power Sources 154:489–496

Munje R, Patre B, Tiwari A (2014) Periodic output feedback for spatial control of AHWR: a three-time-scale approach. IEEE Trans Nucl Sci 61:2373–2382

Munje R, Parkhe J, Patre B (2015a) Control of xenon oscillations in advanced heavy water reactor via two-stage decomposition. Ann Nucl Energy 77:326–334

Munje R, Patil Y, Musmade B, Patre B (2015b). Discrete time sliding mode control for three time scale systems. In: Proceedings of the international conference on industrial instrumentation and control, Pune, 28–30 May 2015, pp 744–749
Munje R, Patre B, Tiwari A (2016) Discrete-time sliding mode spatial control of advanced heavy water reactor. IEEE Trans Control Syst Technol 24:357–364
Na K, Gou B (2008) Feedback linearization based nonlinear control for PEM fuel cells. IEEE Trans Energy Convers 23:179–190
Na K, Gou B, Diong B (2007) Nonlinear control of PEM fuel cells by exact linearization. IEEE Trans Ind Appl 43:1426–1433
Naghidokht A, Elahi A, Ghoranneviss M (2016) Feedback controller design for ignition of deuterium-tritium in NSTX tokamak. Int J Hydrogen Energy 41:15272–15276
Naidu DS (1988) Singular perturbation methodology in control systems. Peter Peregrinus, London
Naidu DS, Calise A (2001) Singular perturbations and time scales in guidance and control of aerospace systems: survey. AIAA J Guid Control Dyn 24:1057–1078
Nazem-Zadeh S, Hamidi-Beheshti MT (2017) Near-optimal controls of a fuel cell coupled with reformer using singular perturbation methods. AUT J Model Simul 49:163–172
Nehrir M, Wang C (2009) Modeling and control of fuel cells: distributed generation applications. Wiley, Hoboken
Nguyen T, White R (1993) A water and heat management model for proton exchange membrane fuel cells. J Electrochem Soc 140:2178–2186
Ogata K (1995) Discrete-time control systems. Prentice Hall, Englewood Cliffs
Onar O, Shirazi O, Khaligh A (2010) Grid interaction operation of a telecommunications power system with a novel topology for multiple-input buck-boost converter. IEEE Trans Power Delivery 25:2633–2645
Ortega J, Reinhardt W (2000) Iterative solution of nonlinear equations on several variables. SIAM, Philadelphia
Padulles J, Ault G, McDonald J (2000) An integrated SOFC plant dynamic model for power systems simulation. J Power Sources 86:495–500
Page S, Anbuky A, Krumdieck S, Brouwer J (2007) Test method and equivalent circuit modeling of a PEM fuel cell in a passive state. IEEE Trans Energy Convers 22:764–773
Pandiyan S, Elayaperumal A, Rajalakshmi N, Dhathathreyan K, Venkateshwaran N (2013) Design and analysis of a proton exchange membrane fuel cells (PEMFC). Renew Energy 49:161–655
Park G, Gajic Z (2012) Sliding mode control of a linearized polymer electrolyte membrane fuel cell. J Power Sources 212:226–232
Park G, Gajic Z (2014) A simple sliding mode controller of a fifth-order nonlinear PEM fuel cell model. IEEE Trans Energy Convers 29:65–71
Pasqualetti F, Dorfler F, Bullo F (2015) Control theoretic methods for cyberphysical security. IEEE Control Syst Mag 35:110–127
Phillips R (1980a) Reduced order modeling and control of two-time scale discrete systems. Int J Control 31:65–780
Phillips R (1980b) Two-stage design of linear feedback controls. IEEE Trans Autom Control 25:1220–1223
Phillips R (1983) The equivalence of time-scale decomposition techniques used in the analysis and design of linear systems. Int J Control 37:1239–1257
Pilloni A, Pisano A, Usai E (2015) Observer based air excess ratio control of a PEM fuel cell system via high order sliding mode. IEEE Trans Ind Electron 6:1–10
Pukrushpan J, Stefanopoulou A, Peng H (2004a) Control of fuel cell power systems: principles, modeling and analysis and feedback design. Springer, London
Pukrushpan J, Peng H, Stefanopoulou A (2004b) Control oriented modeling and analysis for automotive fuel cell systems. Trans ASME J Dyn Syst Meas Control 126:14–25
Pukrushpan J, Stefanopoulou A, Varigonda S, Eborn J, Haugstteter C (2006) Control-oriented model of fuel processor for hydrogen generation in fuel cell applications. Control Eng Pract 14:277–293

Radisavljevic V (2011) On controllability and system constraints of a linear models of proton exchange membrane and solid oxide fuel cells. J Power Sources 196:8549–8552

Radisavljevic V (2013) Optimal parallel controllers and filters for a class of second order linear dynamic systems. J Control Syst Eng 1:37–49

Radisavljevic-Gajic V (2015a) A simplified two-stage design of linear discrete-time feedback controllers. ASME J Dyn Syst Meas Control 138:014506-1–014506-7

Radisavljevic-Gajic V (2015b) Two-stage feedback design for a class of linear discrete-time systems with slow and fast variables. ASME J Dyn Syst Meas Control 138:086502–086507

Radisavljevic-Gajic V (2015c) Full- and reduced-order linear observer implementations in MATLAB/Simulink. IEEE Control Syst Mag 35:91–101

Radisavljevic-Gajic V, Graham K (2017) System analysis of a nonlinear proton exchange membrane fuel cell mathematical model. In: ASME dynamic systems and control conference, Tysons Corner

Radisavljevic-Gajic V, Milanovic M (2016) Three-stage feedback controller design with application to a three-time scale fuel cell system. In: ASME dynamic systems and control conference, Minneapolis

Radisavljevic-Gajic V, Rose P (2014) A new two stage design of feedback controllers for a hydrogen gas reformer. Int J Hydrogen Energy 39:11738–11748

Radisavljevic-Gajic V, Rose P (2015) Optimal observer driven controller with disturbance rejection for a natural gas hydrogen reformer. In: ASME dynamic systems and control conference, 28–30 Oct, Columbus

Radisavljevic-Gajic V, Rose P, Clayton G (2015) Two-stage design of linear feedback controllers for proton exchange membrane fuel cells. In: ASME dynamic systems and control conference, Columbus, doi: https://doi.org/10.1115/DSCC2015-9973

Radisavljevic-Gajic V, Milanovic M, Clayton G (2017) Three-stage feedback controller design with applications to three time-scale control systems. ASME J Dyn Syst Meas Control 139:104502-1–104502-10

Radu R, Taccani R (2006) SIMULINK-FEMLAB Integrated dynamic simulation model for a PEM fuel cell system. ASME J Fuel Cell Sci Technol 6:041015-1

Rao A, Naidu DS (1981) Singularly perturbed difference equations in optimal control problems. Int J Control 34:1163–1174

Reddy B, Samuel P (2017) Technology advancements and trends in development of proton exchange membrane fuel cell hybrid electric vehicles in India: a review. J Green Eng 7:361–384

Rojas A, Lopez G, Gomez-Aguilar J, Alvarado V, Torres C (2017) Control of the air supply system in a PEMFC with balance of plant simulation. Sustainability 9:1–23

Samuelsen S (2017) The automotive future belongs to fuel cells: range, adaptability, and refueling time will ultimately put hydrogen fuel cells ahead of batteries. IEEE Spectr 54:38–43

Sankar K, Jana A (2018a) Dynamics and estimator based nonlinear control of a PEM fuel cell. IEEE Trans Control Syst Technol 26:1124–1131

Sankar K, Jana A (2018b) Nonlinear multivariable sliding mode control of a reversible PEM fuel cell integrated systems. Energ Conver Manage 171:541–565

Sciazko A, Komatsu Y, Brus G, Kimijima S, Szmyd JS (2014) A novel approach to the experimental study on methane/steam reforming kinetics using the Orthogonal Least Squares method. J Power Sources 262:245–254

Seo Y, Seo D, Jeong J, Yoon W (2006) J Power Sources 163:119–124

Serra M, Aguado J, Ansede X, Riera J (2005) Controllability analysis of decentralized linear controllers for polymeric fuel cells. J Power Sources 151:93–102

Severin C, Pischinger S, Ogrzewalla J (2005) Compact gasoline fuel processor for passenger vehicle APU. J Power Sources 145(2):675–682

Shapira I, Ben-Asher J (2004) Singular perturbation analysis of optimal glide. AIAA J Guid Control Dyn 27:915–918

Shimjith S, Tiwari A, Bandyopadhyay B (2011a) Design of fast output sampling controller for three-time-scale systems: application to spatial control of advanced heavy water reactor. IEEE Trans Nucl Sci 58:3305–3316

Shimjith S, Tiwari A, Bandyopadhyay B (2011b) A three-time-scale approach for design of linear state regulators for spatial control of advanced heavy water reactor. IEEE Trans Nucl Sci 58:1264–1276

Simoncini S (2016) Computational methods for linear matrix equations. SIAM Rev 58:377–441

Singh H, Naidu DS, Moore K (1996) Regional pole placement method for discrete-time systems. AIAA J Guid Control 19:974–976

Sinha A (2007) Linear systems: optimal and robust control. Francis & Taylor, Boca Raton

Smith H, Davison E (1972) Design of industrial regulators: integral feedback and feedforward control. Proc IEE 119:1210–1216

Sonntag R, Borgnakke C, Van Wylen G (1998) Fundamentals of thermodynamics. Wiley, New York

Spare B, Iyer V, Lee J, Tice G, Hillman M, Simon D (2011) Fuel cell system to power a portable computing device. Patent application 20110311895, US Patent & Trademark Office

Stefani R, Shahiau B, Savant C, Hostetter G (2002) Design of feedback control systems. Oxford University Press, New York/Oxford

Stewart G (1973) Matrix computations. Academic Press, Orlando, FL, USA

Tong S, Qian D, Fang J, Li H (2013) Integrated modeling and variable fuzzy control of a hydrogen-air fuel cell system. Int J Electrochem Sci 8:3636–3652

Tong S, Fang J, Zhang Y (2017) Output tracking control of a hydrogen-air PEM fuel cell. IEEE/CAA J Automat Sinica 4:273–279

Tsourapas V, Stefanopoulou A, Sun J (2007) Model-based control of an integrated fuel cell and fuel processor with exhaust heat regulation. IEEE Trans Control Syst Technol 15:233–244

Umbria F, Aracil J, Gordillo F (2014) Three-time-scale singular perturbation stability analysis of three-phase power converters. Asian J Control 16:1361–1372

Uzunoglu M, Alam M (2006) Dynamic modeling, design, and simulation of a combined PEM fuel cell and ultracapacitor system for stand-alone residential applications. IEEE Trans Energy Convers 21:767–775

Uzunoglu M, Alam M (2007) Dynamic modeling, design and simulation of a PEM fuel cell/ultra-capacitor hybrid system for vehicular applications. Energ Conver Manage 48:1544–1553

Uzunoglu M, Onar O, El-Sharkh M, Sisworahardjo N, Rahman A, Alam M (2007) Parallel operation characteristics of PEM fuel cell and microturbine power plants. J Power Sources 168:469–476

Wang Z, Ghorbel F (2006) Control of closed kinematic chains using a singularly perturbed dynamics model. Trans ASME J Dyn Syst Meas Control 128:142–151

Wang F-C, Guo Y-F (2015) Robustness analysis of PEMFC systems on the production line. Int J Hydrogen Energy 40:1959–1966

Wang Y-X, Kim Y-B (2014) Real-time control of air excess ratio of a PEM fuel cell system. IEEE/ASME Trans Mechatron 19:852–861

Wang F-C, Peng C-H (2014) The development of an exchangeable PEM power module for electric vehicles. Int J Hydrogen Energy 39:3855–3867

Wang M-S, Li T-H, Sun Y-Y (1996) Design of near-optimal observer-based controllers for singularly perturbed discrete systems. JSME Int J Ser C Dyn Control Robot Design Manuf 39:234–241

Wang G-L, Wang Y, Shi J-H, Shao H-H (2010) Coordinating IMC-PID and adaptive controllers for a PEMFC. ISA Trans 49:87–94

Wang Y, Chen K, Mishler J, Cho S, Adroher X (2011) A review of polymer electrolyte fuel cells: technology, applications and needs on fundamental research. Appl Energy 88:981–1007

Wang F-C, Kuo P-C, Chen H-J (2013) Control design and power management of stationary PEMFC hybrid power system. Int J Hydrogen Energy 38:5845–5856

Wedig W (2014) Multi-time scale dynamics of road vehicles. Probab Eng Mech 37:180–184

Wu X, Zhou B (2016) Fault tolerance control for proton exchange membrane fuel cell systems. J Power Sources 324:804–829

Yalcinoz T, El-Sharkh M, Sisworahardo N, Alam A (2010) Portable PEM fuel cell-ultra capacitor system: model and experimental verification. Int J Energy Res 34:1249–1256

Zenith F, Skogestad S (2009) Control of mass and energy dynamics of polybenzimidazole membrane fuel cells. J Process Control 19:15–432

Zerizer T (2016) Boundary value problem for a three-time-scale singularly perturbed discrete systems. Dynam Contin Discrete Impuls Systems Ser A Math Anal 23:263–272

Zhang Y, Naidu DS, Cai C, Zou Y (2014) Singular perturbation and time scales in control theories and applications: an overview 2002-2012. Int J Inf SystSci 9:1–36

Zhou K, Doyle J (1998) Essential of robust control. Prentice Hall, Upper Saddle River

Zhou D, Al-Dura A, Gao F, Ravey A, Matraji I, Simoes M (2017) Online energy management strategy of fuel cell hybrid electric vehicles based on data fusion approach. J Power Sources 366:278–291

Zhou D, Al-Dura A, Matraji I, Ravey A, Gao F (2018) Online energy management strategy of fuel cell hybrid electric vehicles: a fractional-order extremum seeking method. IEEE Trans Ind Electron 65:6787–6799

Zhu Q, Basar T (2015) Game-theoretic methods for robustness, security, and resilience of cyberphysical control systems. IEEE Control Syst Mag 35:46–65

Zhu Y, Tomsovic K (2002) Development of models for analyzing the load-following performance of micro-turbines and fuel cells. Electr Pow Syst Res 62:1–11

Zhu J, Zhang D, King K (2001) Reforming of CH_4 by partial oxidation: thermodynamic and kinetic analysis. Fuel 80:899–905

zur Megede D (2002) Fuel cell processors for fuel cell vehicles. J Power Sources 106:35–41

Index

V. Radisavljevic-Gajic et al., *Multi-Stage and Multi-Time Scale Feedback Control of Linear Systems with Applications to Fuel Cells*, Mechanical Engineering Series,
https://doi.org/10.1007/978-3-030-10389-7

Printed by Printforce, the Netherlands